Lecture Notes in Statistics

Edited by D. Brillinger, S. Fienberg, J. Gani, J. Hartigan, and K. Krickeberg

39

James D. Malley

Optimal Unbiased Estimation of Variance Components

Springer-Verlag

Berlin Heidelberg New York London Paris Tokyo

Author

James D. Malley
National Institutes of Health
Division of Computer Research and Technology
Laboratory of Statistical and Mathematical Methodology
Building 12A
Bethesda, Maryland 20892, USA

AMS Subject Classification (1980): 17C20, 17C50, 62F10, 62J10

ISBN 3-540-96449-5 Springer-Verlag Berlin Heidelberg New York
ISBN 0-387-96449-5 Springer-Verlag New York Berlin Heidelberg

Printed in Germany

Printing and binding: Druckhaus Beltz, Hemsbach/Bergstr.
2147/3140-543210

MS-SHK
SCIMON

Introduction

The clearest way into the Universe is through a forest wilderness.

John Muir

As recently as 1970 the problem of obtaining optimal estimates for variance components in a mixed linear model with unbalanced data was considered a miasma of competing, generally weakly motivated estimators, with few firm guidelines and many simple, compelling but unanswered questions.

Then in 1971 two significant beachheads were secured: the results of Rao [1971a, 1971b] and his MINQUE estimators, and related to these but not originally derived from them, the results of Seely [1971] obtained as part of his introduction of the notion of quadratic subspace into the literature of variance component estimation.

These two approaches were ultimately shown to be intimately related by Pukelsheim [1976], who used a linear model for the components given by Mitra [1970], and in so doing, provided a mathematical framework for estimation which permitted the immediate application of many of the familiar Gauss-Markov results, methods which had earlier been so successful in the estimation of the parameters in a linear model with only fixed effects. Moreover, this usually enormous linear model for the components can be displayed as the starting point for many of the popular variance component estimation techniques, thereby unifying the subject in addition to generating answers.

Continuing with our history, the classical maximum likelihood approach to the problem of estimating the components moved along, although slowly, and the ML equations still remain very difficult to solve even for the simplest mixed-model balanced data situations. And this numerical intractability seems to haunt other methods of estimation which are enviably free of serious theoretical defects.

More recently, the author has brought a part of the theory of unbiased estimation to its logical completion, by finding simple, necessary and sufficient conditions for an estimable function of the parameters to have a best quadratic unbiased translation invariant estimate, for any mixed model with unbalanced data and arbitrary kurtosis. These conditions were found by building on the Pukelsheim-Mitra model and by relying on the elegant and broadly useful Lehmann-Scheffé result on minimum variance unbiased estimation. The conditions for the existence of such estimates are rather algebraic in character, which leads naturally to the study of the ring-theoretic structure of the estimation problem and its practical consequences.

The variance component estimation problem, for the general mixed model, with unbalanced data and arbitrary kurtosis, may now be described as having two broad streams. The first is that of unbiased estimation, which is, with the results presented here, essentially complete. Unfortunately, while the unbiased theory leads to estimates which are optimal, and relatively easy to calculate, it has the persistent problem of generating estimates which may be strictly negative with positive probability. The second stream is that of biased estimation, and here we find the reverse situation: while mostly free of theoretical defect, the estimates are also usually free of theoretical optimality and, further, the results are primarily of the existence type only, with explicit estimators presently available in only the simplest of models.

While it seems safe to assume that this numerical impasse will yield in time to the increased powers of new hardware and computing techniques, we choose to elaborate only the unbiased theory which has now been brought to completion.

We do this also because the techniques introduced here deserve a wider audience and a more general application in statistics. These methods are more ring-theoretic and algebraic in nature than generally encountered in statistics. They also represent a logical progression of technique that has seen its fullest expression in the linear methods described in Rao's Linear Statistical Inference and its Applications [1973a] and more recently, in significant portions of Farrell's Techniques of Multivariate Calculation [1976; 1985]. Some techniques used also appear in the design studies of James [1957], Mann [1960], Ogawa and Ishii [1965], and Robinson [1970, 1971].

The advance is thus from essentially vector space methods, with some call upon convexity arguments, to ring-type manipulations and the introduction of algebras and some structure theory. Using algebraic methods greatly simplifies the problem of variance component estimation, and it is the author's view that such methods have an even broader utility in statistics than that limited to this study.

Since a ring-theoretic approach to the estimation problem invokes ideas and constructions which may be unfamiliar, we have included background material on rings, ideals and algebras, both associative and non-associative. Hopefully this material will save the reader from having to track unescorted through the algebra literature.

A new formulation of the Gauss-Markov theorem has also been included, based on work of Zyskind [1967], Rao [1967], and others. While this latter material is not needed for the results we obtain on unbiased estimation of variance components, it enables our discussion to be placed in context. Its inclusion is also used as an opportunity to set aright and fully explicate certain steps in this circle of ideas that have been evidently absent in the literature.

We will also be making reference to a larger context on other occasions, and this will result in a dual approach to some proofs, in that a shorter derivation along more familiar lines will be given together with one that is a specialization of a general result.

Examples of data analysis, additional representative models for variance components analysis, and a more complete historical discussion can be found elsewhere in the statistical literature, for example in S. R. Searle's Linear Models [1971], his Cornell University Biometrics Unit Notes [1979], and in Kleefe [1977].

An extensive bibliography can be found in Sahai [1979] and Sahai, Khuri, and Kapadia [1985]. Also, a very recent and complete overview of the last thirty years of variance component estimation is found in Khuri and Sahai [1986].

We use the decimal system for numbering theorems, lemmas and corollaries, and number consecutively, so that the second such item in Section 6 in Chapter 1 say, is Lemma 1.6.2.

For generously taking time to read and comment on early versions of the manuscript I would like to thank George Hutchinson, George Casella, and Gregory Campbell.

Finally, I wish to thank my wife Karen, as well as Linda Hirsch and James Mosimann for their warm support and encouragement of this project. I also want to particularly thank my wife for her extensive and insightful editorial assistance.

James D. Malley
Bethesda, Maryland
May 12, 1986

Table of Contents

Chapter One: The Basic Model and The Estimation Problem

1.1 Introduction.

Here's what we're going to do: starting with a simple example of a variance component model, we move to the matrix formulation of the general mixed model. Then we state the essential estimation problem. Relationships among a constellation of possible estimation criteria are next studied, and finally an ordered combination of these key ideas is taken as the focus of our unbiased theory. In Chapter Ten we will look at a new combination of criteria involving non-negativity.

1.2 An Example.

Consider the following biomedical problem: mice are used as an animal model to study the nature and progress of an infection by a parasitic worm of the genus Schistosoma. Equal numbers of cercariae (the infecting intermediate stage of the parasite) are introduced into the mice, and after eight weeks infection intensity is measured by counting the number of worm-pairs (male-female) in the liver of each mouse.

The experiment uses three strains of mice and two strains or subspecies of schistosomes, and it is assumed that the number of infecting cercariae given each mouse is not so great as to kill the host within the eight week trial.

It is of interest to examine the variability of the susceptibility to infection for each mouse strain and each worm strain. The basic variable for the analysis is y_{ijk} = log (worm-pair total) for each mouse, expressed in the following mixed model:

$y_{ijk} = \mu + \alpha_i + b_j + c_{ij} + e_{ijk}$ where

μ = constant

i = mouse strain, i = 1, 2, 3.

j = worm strain, j = 1, 2.

k = mouse number

α_i = ith fixed effect for the mouse strain

b_j = jth random effect for the worm strain

c_{ij} = random interaction effect between mouse-worm strain

e_{ijk} = measurement error.

We assume the the random variables (r.v.'s) b_j, c_{ij}, and e_{ijk} have means, variances and covariances as follows:

$E(b_j) = 0$; $E(c_{ij}) = 0$; $E(e_{ijk}) = 0$

$var(b_j) = \sigma_b^2 \geq 0$ all j; $var(c_{ij}) = \sigma^2 \geq 0$ all i, j;

$var(e_{ijk}) = \sigma_e^2 \geq 0$ all i, j, k.

$cov(b_j, b_k) = 0$, $j \neq k$; $cov(c_{ij}, c_{rs}) = 0$,
unless i = r, j = s;

$cov(b_j, c_{rs}) = 0$;
$cov(b_j, e_{rst}) = 0$;
$cov(c_{ij}, e_{rst}) = 0$
for all i, j, r, s, t.

We wish to estimate σ_b^2, σ_c^2, and σ_e^2 and make comparisons between these three values.

It is also of interest to examine the resistance of the host to subsequent infections after the initial infection. Here, mice are given a second infection ten weeks after an initial one, and the livers are examined four weeks later. We choose to express resistance to the second infection as the ratio

$$\log \left[\frac{\text{total worm-pairs from second infection}}{\text{total control infection}} \right]$$

where the denominator is the mean of the worm-pair totals for a group of mice of the same strain as the given mouse, but having only the second infection of the same worm strain, and no initial infection. Once more we want to estimate the variances in the model above, where now the data vector is the resistance ratio.

Other examples, as well as a summary of what was known about variance component estimation up until 1971 can be found in Searle [1971]. The example above is typical of variance component estimation problems in that it has these features:

[1] Ratios of functions of the components are usually the targets of analysis.

[2] Exact or even approximate hypothesis testing is not mentioned. This state of affairs is presently unavoidable, since for only the most elementary functions of the component estimates are the distribution problems well understood, even for originally normal data. Progress is slowly being made, though -- see Seely and El-Bassiouni [1983], and Morin-Wahhab [1985].

[3] Little attention is given to the fixed-effects part of the model. Indeed, it is usually thought desirable to use estimates of the components which are left unchanged under a change of the fixed-effects:

$$\mu \to \mu^*, \; \alpha_i \to \alpha_i^*.$$

However, the question of obtaining estimates of the

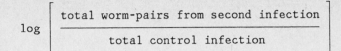

fixed-effects in the presence of the variance components has been studied. For example, recently Searle [1984] has shown how minimum variance estimates of the fixed effects can be found for all balanced data mixed models. See also Kleefe [1978].

1.3 The Matrix Formulation.

We now shift to the matrix formulation of the general mixed model, since without using matrices the notation is prohibitively subscript-intensive.

Begin with a random n-vector of data, $y \in \mathbf{R}_n$ and assume that

$$E(y) = X\alpha + \sum_{i=1}^{i=k} Z_i b_i \qquad [1]$$

where α is the fixed-effects parameter vector, $\alpha \in \mathbf{R}_t$
each b_i is a random c_i-vector, $b_i \in \mathbf{R}_{c_i}$
X, $n \times t$, is the fixed-effects part of the mixed model
Z_i, $n \times c_i$, is the design matrix associated with the ith random part b_i.

Form the partitioned matrix

$$Z = (Z_1 \mid Z_2 \mid \dots \mid Z_k)$$

where it is assumed that $Z_k = I_n$, the $n \times n$ identity matrix.

Next, take

$$\operatorname{var}(y) = V = ZDZ' \qquad [2]$$

for D a diagonal matrix of the variances of the components of each b_i, so that

$$\operatorname{var}(b_i) = \sigma_i^2 I_{c_i}, \quad \text{with} \sum_{i=1}^{i=k} c_i = m.$$

and

$$D = \bigoplus_{i=1}^{i=k} \sigma_i^2 I_{c_i} = \text{the block diagonal matrix with}$$

$$\text{ith block} = \sigma_i^2 I_{c_i} .$$

Next, let

$$V_i = Z_i Z_i' \quad \text{for all } i,\ 1 \le i \le k$$

and write [2] as

$$\text{var}(y) = V = \sum_{i=1}^{i=k} \sigma_i^2 V_i .$$

Furthermore, let

$$\sigma^2 = (\sigma_1^2 , \sigma_2^2 , \ldots, \sigma_k^2)$$

$$\gamma = (\gamma_1 , \gamma_2 , \ldots, \gamma_k)$$

where we write γ_i for the common kurtosis parameter of each component of the r.v. b_i so that

$$\gamma_i = [\mu_i / (\sigma_i^2)^2] - 3$$

with μ_i the common fourth central moment of each component of b_i.

Equations [1] and [2] constitute our basic mixed model. We make no further distributional assumptions except to assume, of course, that the first four moments of the r.v. y exist.

1.4 Estimation Criteria.

We list a number of properties of an estimate g(y) that have been considered worthwhile criteria for estimation of the components σ_i^2:

[1] unbiasedness: given a parametric function $h(\sigma^2)$, we must have

$$E[g(y)] = h(\sigma^2), \text{ for all } \sigma^2.$$

[2] translation invariance: for any t-vector δ, $\delta \in \mathbf{R}_t$, if we let the r.v. y_δ have expectation $X(\alpha + \delta)$, with $var(y_\delta) = var(y)$ then

$$g(y_\delta) = g(y).$$

[3] quadratic: we take $g(y) = y'Ay$ for some $n \times n$ symmetric matrix A.

[4] minimum variance: among a given class of functions $g(y)$ we want the one having minimum variance, if such a function exists.

[5] minimum norm: for a given positive semidefinite matrix W, $n \times n$, we use the quadratic estimator $g(y) = y'Ay$ having $trace(AWA)$ a minimum.

[6] non-negativity: we require that the function $g(y)$ be non-negative, so that $g(y) \geq 0$, for all realizations of the r.v. y.

[7] minimum bias: for all $g(y)$ which are estimates of $h(\sigma^2)$, we select the one with smallest bias, so that

$$\{E[g(y) - h(\sigma^2)]\}^2$$

is a minimum.

[8] minimum bias in norm: for $h(\sigma^2) = \Sigma p_i\sigma_i^2$, we select that estimate $g(y)$ such that $E(g(y)) = \Sigma g_i\sigma_i^2$ implies $\Sigma [g_i - p_i]^2$ is a minimum.

The list does not exhaust the collection of all properties considered useful in one context or another, but does include all those we will be studying here.

1.5 Properties of the Criteria.

We next consider some consequences and requirements of the criteria proposed above. We write tr(A) for trace(A).

[1] If we require of g(y) that it be unbiased and quadratic then $g(y) = y'Ay$ for some symmetric A. We can now use a standard result on the expectations of quadratic forms, valid for all r.v.'s having finite second moments:

$$E[g(y)] = E[y'Ay] = E[tr(y'Ay)] = E[tr(Ayy')]$$

$$= tr[E(Ayy')] = tr[AE(yy')]$$

$$= trA[var(y) + (X\alpha)(X\alpha)'] = tr(AV) + \alpha'X'AX\alpha$$

From unbiasedness we must have $E[y'Ay] = h(\sigma^2)$, say, for all σ^2 and all α. But since

$$V = \sum_{i=1}^{i=k} \sigma_i^2 V_i$$

and since unbiasedness must hold for $\alpha = 0$ in particular, then

$$h(\sigma^2) = tr(AV) = tr(A \sum_{i=1}^{i=k} \sigma_i^2 V_i) = \sum_{i=1}^{i=k} \sigma_i^2 tr(AV_i).$$

Moreover, if $h(\sigma^2)$ is linear in the components σ_i^2, so that

$$h(\sigma^2) = \sum_{i=1}^{i=k} p_i \sigma_i^2 ,$$

for some k-vector $p \varepsilon \mathbb{R}_k$, then

$$p_i = tr(AV_i) \text{ for all } i, \ 1 \le i \le k.$$

Lemma 1.5.1. [a] The quadratic estimate $y'Ay$ is unbiased for $p'\sigma^2$, with $p' = (p_1, p_2, \ldots, p_k)$, if and only if $X'AX = 0$ and

$p_i = \mathrm{tr}(AV_i)$ for all i;

[b] If y'Ay is translation invariant, then X'AX = 0, and if in addition, y assumes all members of a basis for R_n, then AX = 0.

Proof. Part [a] has been shown above, while both parts of [b] follow from consideration of the invariance condition for all α and y:

$$y'Ay = (y - X\alpha)'A(y - X\alpha)$$

$$= y'Ay - \alpha'X'Ay - y'X'A\alpha + \alpha'X'AX\alpha. \quad \blacksquare$$

We will not generally assume that y satisfies the additional condition of [b], since var(y) is not necessarily positive definite, so instead let's look more closely at the condition X'AX = 0.

Let $\mathcal{P}_X$ be the projection operator in $\mathbb{R}_n$ onto the subspace spanned by the columns of X and write $M = I_n - \mathcal{P}_X$. Using Rao and Mitra [1971, p. 24] results in the following:

Lemma 1.5.2. For A real and symmetric

$$X'AX = 0 \iff A = B - \mathcal{P}_X B \mathcal{P}_X \quad \text{for some B real and symmetric}$$

$$\iff A = MBM + \mathcal{P}_X B + B\mathcal{P}_X - 2\mathcal{P}_X B \mathcal{P}_X.$$

Corollary 1.5.3. A sufficient, but not necessary, condition for A to be such that X'AX = 0 is that A = MCM for some real, symmetric C.

Lemma 1.5.4. For A p.s.d., X'AX = 0 $\iff$ A = MBM for some real and symmetric B.

[2] Let's now assume that g(y) is quadratic and non-negative. We must have

$$g(y) = y'Ay \geq 0 \text{ for all realizations of the r.v. y.}$$

This is not exactly the same as requiring that the matrix A be positive semidefinite since we generally do not assume that the r.v. y assumes all real n-vectors. In particular V may be singular, and as we shall see later, this implies that with probability one, the r.v. y is restricted to a proper subspace of R_n. Nevertheless, we will take the requirement of non-negativity of the estimate to mean non-negativity of the matrix A, that is, that A be positive semi-definite.

Finally, pulling all the above together gives:

Lemma 1.5.5. If A = MBM then y'Ay is translation invariant. If y'Ay is unbiased for $p'\sigma^2$ then A has the form given in Lemma 1.5.2. If y'Ay is non-negative and unbiased for $p'\sigma^2$ then A = MBM.

Additional connecting links between these conditions can be found in LaMotte [1973b].

1.6 Selection of Estimation Criteria.

We now choose our principle criteria for estimation. We want our estimates to be:

[a] quadratic,

[b] unbiased,

[c] translation invariant.

We will assume from now on that all estimates we consider will satisfy [c], and so be translation invariant. We will call an estimate optimal if either

[d] they have minimum variance in the class of all estimates satisfying [a], [b] and [c]; in the literature

such estimates are called best quadratic translation invariant (invariant BQUE), while we will simply call them <u>optimal unbiased</u>;

OR

[e] they have minimum variance in the class of all estimates satisfying [b], and [c], in which case they are the uniform minimum variance unbiased translation invariant estimates, and we write <u>invariant UMVUE</u>, or just <u>I-UMVUE</u>.

Note that from the preceding lemmas it is sufficient to start with y'Ay for A = MBM in order to satisfy [a] and [c]. Then y'Ay = (My)'A(My) and the model is transformed to

$$y_* = My, \qquad E(y_*) = 0, \qquad \mathrm{var}(y_*) = V_* = MVM.$$

We will refer to this as the <u>basic reduced model</u> and it will be used as the starting point for estimation. Letting $M_i = MZ_iZ_i'M$ we express the basic reduced model as:

$$E(y_*) = 0, \quad \mathrm{var}(y_*) = V_* = \sum_{i=1}^{i=k} \sigma_i^2 M_i, \quad \text{with } M_k = M = I_n - \mathcal{P}_X.$$

Chapter Two: Basic Linear Technique

2.1 Introduction.

In this chapter some linear and matrix methods are introduced that will simplify later work, particularly the linearization of the basic reduced model, a subject we take up in the next chapter. The notions of vec, $\text{mat}_{(m,n)}$, $I_{(m,n)}$, tensor products $\otimes$, and some relations between them are discussed. Next are the space of symmetric matrices, its natural inner product, and a useful projection lemma.

A detailed discussion of vec and $I_{(m,n)}$ and their relationship to the tensor product $\otimes$ can be found in Henderson and Searle [1979]. Other sources for the material in this chapter are the texts Graham [1981] and Rogers [1980].

2.2 The vec and mat operators.

For many of our technical arguments later, it will be most convenient if we work just with vectors instead of matrices, or conversely. Thus the general linear model is usually thought of and analysed in an essentially vector context ("begin with a vector of observations"), yet here as elsewhere the vector/matrix selection is largely one of tradition and is only occasionally for conceptual simplicity. In order then to make a choice for simplicity we have need of a mechanism for conveniently making this vector/matrix conversion.

Our basic linearizing operator is vec(·), which is simply, for a given matrix A, $n \times m$, vec(A) = the vector $\varepsilon\ \mathbb{R}_{nm}$ formed by stacking the columns of A one upon the other, beginning with column 1, moving left to right. Thus

$$A = \begin{bmatrix} 1 & 2 \\ 3 & 4 \end{bmatrix} \qquad \text{has } \mathrm{vec}(A) = (1,\ 3,\ 2,\ 4)'.$$

Next we introduce the inverse of vec(·), an idea evidently new to the literature. Thus for a vector $v \in \mathbb{R}_k$ and a given factorization of k as k = nm, we will write $\mathrm{mat}_{n,m}(v) \in \mathbb{R}_{n,m}$ for the n × m matrix formed by segmenting v into m subvectors each of length n and placing them in an array, left to right. Thus

$$v = (1,\ 2,\ 3,\ 4,\ 5,\ 6)'$$

has

$$\mathrm{mat}_{2,3}(v) = \begin{bmatrix} 1 & 3 & 5 \\ 2 & 4 & 6 \end{bmatrix}$$

while

$$\mathrm{mat}_{3,2}(v) = \begin{bmatrix} 1 & 4 \\ 2 & 5 \\ 3 & 6 \end{bmatrix}$$

We will write $\mathrm{mat}_n(\cdot)$ for $\mathrm{mat}_{n,n}(\cdot)$ and drop the indexing altogether if it is is clear from the context.

Note that the vectors vec(A) and vec(A') contain the same elements, though in a different order: the matrix operator connecting these vectors is the vec-permutation matrix, also called the permuted identity matrix, $I_{(m,n)}$. This is the square mn × mn matrix partitioned into submatrices all of order m × n such that the i,jth submatrix has j,ith element = 1, with zeroes elsewhere. $I_{(m,n)}$ is operationally defined from

$$\mathrm{vec}(A) = I_{(m,n)}\ \mathrm{vec}(A').$$

As an example, $I_{(2,3)}$ is

$$\begin{bmatrix} 1\,0\,0 & 0\,0\,0 \\ 0\,0\,0 & 1\,0\,0 \\ 0\,1\,0 & 0\,0\,0 \\ 0\,0\,0 & 0\,1\,0 \\ 0\,0\,1 & 0\,0\,0 \\ 0\,0\,0 & 0\,0\,1 \end{bmatrix}$$

Next we define the tensor product, $\otimes$, of two matrices, A $r \times s$, B $k \times t$:

$A \otimes B = (a_{ij}B)$, with a_{ij} the i,j-th element of A and

$a_{ij}B$ the i,j-th block of size $k \times t$.

2.3 Useful Properties of the Operators.

We pull these definitions together with a collection of relations, for which a good general reference is Henderson and Searle [1979]. Here, A is $r \times s$, B is $k \times t$.

[1] $B \otimes A = I_{(r,k)}\ (A \otimes B)\ I_{(t,s)}$

[2] $\mathrm{vec}(ABC) = (C' \otimes A)\ \mathrm{vec}(B)$

[3] for column vectors $a \in \mathbb{R}_m$, $b \in \mathbb{R}_n$,

$I_{(m,n)}(a \otimes b) = b \otimes a$

[4] row (i, j) of $I_{(m,n)}$ is row (j, i) of I_{mn}

[5] letting e_i be the i-th unit vector in $\mathbb{R}_m$, and e_j be the unit vector in $\mathbb{R}_n$

$$I_{(m,n)} = \sum_{i=1}^{i=m} (e_i \otimes I_n \otimes e_i)$$

$$= \sum_{j=1}^{j=n} (e_j \otimes I_m \otimes e_j)$$

[6] $[I_{(m,n)}]\ [I_{(n,m)}] = I_{mn}$; $\quad I_{(m,1)} = I_{(1,m)} = I_m$

[7] $I_{(r,k)} (A \otimes B) = (B \otimes A) I_{(s,t)}$

[8] $I_{(mp,s)} = I_{(pm,s)}$
$= [I_{(m,ps)}] [I_{(p,ms)}]$
$= [I_{(p,s)} \otimes I_m] [I_p \otimes I_{(m,s)}]$

where I_m is the $m \times m$ identity matrix

[9] $tr(AB) = (vec\ A')' (vec\ B)$

[10] for vectors $a \in \mathbb{R}_n$, $b \in \mathbb{R}_m$, $vec(ab') = a \otimes b$

[11] for symmetric A, $r \times r$: $vec\ A = I_{(r,r)} [vec\ A]$

Other properties of the tensor product can be found in Rao and Mitra [1971], and include

[12] $(A + B) \otimes C = (A \otimes C) + (B \otimes C)$

[13] $A \otimes (B + C) = (A \otimes B) + (A \otimes C)$

[14] $(A \otimes B) (C \otimes D) = AC \otimes BD$

[15] $(A \otimes B)^- = A^- \otimes B^-$, for g-inverses A^-, B^- of A and B

[16] for the partitioned matrix $(A \mid B)$: $(A \mid B) \otimes C = (A \otimes C \mid B \otimes C)$. But note that tensoring from the other side does not push through in general: $A \otimes (B \mid C) \neq (A \otimes B \mid A \otimes C)$

[17] $(A \otimes B)' = A' \otimes B'$

Chapter Three: Linearization of the Basic Model

3.1 Introduction.

In this chapter we use the pioneering work of Pukelsheim and Mitra to transform the model into a simply linear one. This has the double benefit of better organization for the statement of the model, and of allowing us to apply where possible the large body of technique available for the linear model. We also present an alternative linearization of the basic model.

Henceforward we will drop indices on our summations and assume the natural range will be from 1 to k, except as noted.

3.2 The First Linearization.

Pukelsheim [1976], in a different notation, argued as follows.

With $M = I_n - \mathcal{P}_X$ as usual, begin with the basic reduced model of Section 1.6:

$$\text{vec }[\text{var}(My)] = \text{vec }[M\text{var}(y)M] = \text{vec } M[\ \Sigma\ \sigma_i^2 V_i]M.$$

Next recall that $M_i = MZ_iZ_i'M = MV_iM$, so

$$\begin{aligned}\text{vec }[\text{var}(My)] &= \text{vec }[\ \Sigma\ \sigma_i^2 M_i] = \Sigma\ \sigma_i^2[\text{vec}M_i] \\ &= [\text{vec } M_1 \mid \text{vec } M_2 \mid \dots \mid \text{vec } M_k]\ \sigma^2.\end{aligned}$$

Since $E(My) = MX\alpha = 0$ we see

$$\begin{aligned}E[\text{vec }[(My)(My)']\ ] &= \text{vec } E[(My)(My)'] \\ &= \text{vec }[\text{var}(My)].\end{aligned}$$

Now let

$$\mathscr{y} = My \otimes My = vec\ [(My)(My)'].$$

Then

$$E[\mathscr{y}] = E[vec\ [(My)(My)']\ ]$$

$$= [vec\ M_1\ |\ vec\ M_2\ |\ \ldots\ |\ vec\ M_k]\ \sigma^2.$$

Next put

$$\mathscr{X}_* = [vec\ M_1\ |\ vec\ M_2\ |\ \ldots\ |\ vec\ M_k].$$

Bringing the above together:

[1] $$E[\mathscr{y}] = \mathscr{X}_* \sigma^2.$$

Equation [1] is the fundamental linearization of our original variance components model. It is linear in the components σ_i^2, and as it stands makes no distributional assumptions on the data vector y or the constructed r.v. $\mathscr{y}$. The model is sometimes referred to as the "dispersion-mean" model, and its centrality in a group of the common variance component estimation methods will be demonstrated shortly.

Before proceeding to calculate the covariance matrix of $\mathscr{y}$, and so completing the specification of our linear model, we make the following observations.

In writing down [1] and then moving on to use the body of linear model technique, we are implicitly allowing the components in the model to take on all real values, positive and negative. In a sense this is an enlargement of the parameter space and is a hidden assumption which violates the physical reality of the components, since they are variances of r.v.'s, hence non-negative. This expansion of the parameter space may allow the application of certain powerful techniques, but also runs the risk of ending with estimates which are negative, with positive probability. Searle [1971, p. 406] gives such an example. We conjecture here that for every model it is actually possible to construct simple data sets having this probability arbitrarily close to one.

This problem is in fact generally pervasive among variance component estimation procedures, beyond any of those just based on our linear model, and has been for many years a source of much chagrin and aggravation among statisticians. Of course there are methods which detour around this problem at the outset, maximum likelihood being a prominent example, but we choose for now to leave the issue identified but unresolved, at least as far as unbiased estimation is concerned, returning to it only in the last chapters after having built up an algebraic structure theory for the components.

3.3 Calculation of var(y).

We follow the derivation of Anderson [1978] in his thesis (Chapter 4), and calculate the variance of $(y - X\alpha) \otimes (y - X\alpha)$. Let's write

$$F = \text{var}[(y - X\alpha) \otimes (y - X\alpha)],$$

so that

$$F = \text{var}[(Zb) \otimes (Zb)] = \text{var}[(Z \otimes Z)(b \otimes b)]$$

$$= (Z \otimes Z)(\text{var}[b \otimes b])(Z \otimes Z)'.$$

where we recall that $E(b) = 0$. Now

$$\text{var}(b \otimes b) = E[(b \otimes b)(b \otimes b)'] - [E(b \otimes b)]\,[E(b \otimes b)]'.$$

Using the vec(·) property, $\text{vec}\,(ab') = a \otimes b$, we get

$$\text{var}(b \otimes b) = E[(bb') \otimes (bb')] - E[\text{vec}(bb')]\,E[\text{vec}(bb')]'$$

$$= E[(bb') \otimes (bb')] \quad - \quad (\text{vec}\,[E(bb')])\,(\text{vec}\,[E(bb')])'$$

$$= E[(bb') \otimes (bb')] \quad - \quad (\text{vec}\,D)(\text{vec}\,D)'.$$

Needed then is an expression for the first term above, which is the matrix of all the fourth moments of the r.v. $b \in \mathbf{R}_m$.

For this, let's assume now that matrix D is positive definite, with no zero components. In our derivation this results in no loss of generality since if a variance is zero, all higher moments are zero, and the term can be dropped here. Thus we can write

$$w = D^{-\frac{1}{2}}b,$$

so that w has $m = \Sigma\, c_i$ components. This new r.v. has the properties

$$E(w) = 0, \quad \text{and} \quad E(ww') = \text{var}(w) = I_m.$$

Also

$$E(w^4) = 3 + \dot{\gamma}_j, \quad \text{for } j = 1, \ldots, m,$$

where

$$\dot{\gamma}_j = \text{the jth diagonal element of } \ (\oplus\, \gamma_i I_{c_i}).$$

Consequently

$$E[(bb') \otimes (bb')] = (D^{\frac{1}{2}} \otimes D^{\frac{1}{2}})\, E[(ww') \otimes (ww')]\, (D^{\frac{1}{2}} \otimes D^{\frac{1}{2}})$$

$$= (D^{\frac{1}{2}} \otimes D^{\frac{1}{2}})\, \Sigma\, (D^{\frac{1}{2}} \otimes D^{\frac{1}{2}}),$$

where we define

$$\Sigma,\ m \times m,\ = \{\Sigma_{j\ell}\}, \text{ for } j, \ell = 1, \ldots, k,$$

$$= \{E(w_j w_\ell w_r w_s)\}, \text{ for } j, \ell, r, s = 1, \ldots, m.$$

Next, let's evaluate Σ. For $j = \ell$,

$$E(w_j w_\ell w_r w_s) = \begin{cases} 3 + \dot{\gamma}_j & \text{when } j = r = s \\ 1 & \text{when } j \neq r = s \\ 0 & \text{otherwise.} \end{cases}$$

and for $j \neq \ell$

$$E(w_j w_\ell w_r w_s) = \begin{cases} 1 & \text{when } j = r, \ \ell = s \\ 1 & \text{when } j = s, \ \ell = r \\ 0 & \text{otherwise.} \end{cases}$$

Now in $\mathbb{R}_m$ let e_j be the jth unit vector, so

$$\Sigma_{jj} = I_m + (2 + \dot{\gamma}_j) e_j e_j'$$

and

$$\Sigma_{j\ell} = e_j e_\ell' + e_\ell e_j', \quad \text{for } j \neq \ell.$$

Since with $j = \ell$ we have

$$2 e_j e_j' = e_j e_\ell' + e_\ell e_j'$$

so we get

$$\Sigma = I_{m^2} + \{e_j e_\ell' + e_\ell e_j'\} + (\oplus \dot{\gamma}_j e_j e_j').$$

where the indices in the middle term run from 1 to m, and where the matrix in question is composed of submatrices, whose (j, ℓ)th block is $e_j e_\ell' + e_\ell e_j'$.

One next checks that

[1] $\{e_\ell e_j'\} = I_{(m,m)}$;

[2] $\{e_j e_\ell'\} = (\text{vec } I)(\text{vec } I)'$;

[3] $\oplus \dot{\gamma}_j \{e_j e_j'\} = \text{diag}[\text{vec}(\oplus \gamma_i I_{c_i})]$.

Using the above leads to

$$\Sigma = I_{m^2} + I_{(m,m)} + (\text{vec } I_m)(\text{vec } I_m)' + \text{diag}[\text{vec}(\oplus \gamma_i I_{c_i})].$$

Thus

$$E[(bb') \otimes (bb')] =$$

$$(D^{\frac{1}{2}} \otimes D^{\frac{1}{2}})[I_{m^2} + I_{(m,m)} + (\text{vec } I_m)(\text{vec } I_m)' + G](D^{\frac{1}{2}} \otimes D^{\frac{1}{2}}),$$

where we have written

$$G = \text{diag } [\text{vec } (\oplus \gamma_i I_{c_i})].$$

Using [7] from 2.3 gives

$$(D^{\frac{1}{2}} \otimes D^{\frac{1}{2}}) \, I_{(m,m)} \, (D^{\frac{1}{2}} \otimes D^{\frac{1}{2}}) = (D \otimes D) \, I_{(m,m)},$$

and using [2] from 2.3

$$(D^{\frac{1}{2}} \otimes D^{\frac{1}{2}}) \text{ vec } I_{m^2} = \text{vec } D,$$

and

$$(\text{vec } I_{m^2})' \, (D^{\frac{1}{2}} \otimes D^{\frac{1}{2}}) = (\text{vec } D)'.$$

Also, the following notation

$(D^{\frac{1}{2}} \otimes D^{\frac{1}{2}}) \, G \, (D^{\frac{1}{2}} \otimes D^{\frac{1}{2}}) = \Gamma$, say, yields

$$\Gamma = \text{diag } [\text{vec } (\oplus \gamma_i \sigma_i^4 I_{c_i})].$$

This all leads to

$$E[(bb') \otimes (bb')] = (D \otimes D)[I_{m^2} + I_{(m,m)}] + (\text{vec } D)(\text{vec } D)' + \Gamma.$$

Referring back to our original expression for var $(b \otimes b)$ we now see

$$\text{var } (b \otimes b) = (D \otimes D)(I_{m^2} + I_{(m,m)}) + \Gamma.$$

Hence finally,

$$F = (Z \otimes Z) \, [(D \otimes D)(I_{m^2} + I_{(m,m)}) + \Gamma] \, (Z \otimes Z)'.$$

And since $ZDZ' = V$,

$$F = (V \otimes V)(I_{n^2} + I_{(n,n)}) + (Z \otimes Z)\Gamma(Z \otimes Z)',$$

where we have used [7] again from 2.3. That is,

$$I_{(m,m)} (Z \otimes Z)' = (Z \otimes Z)' I_{(n,n)},$$

as Z is $n \times m$, and

$$(Z \otimes Z)(D \otimes D)(Z \otimes Z)' = (ZDZ') \otimes (ZDZ') = V \otimes V.$$

This expression for F is the desired general variance we sought. Next, since $M(y - X\alpha) = My$, and as

$$\begin{aligned} var[(My)(My)'] &= var[M(y - X\alpha)(y - X\alpha)'M] \\ &= M[var(y - X\alpha)(y - X\alpha)']M, \end{aligned}$$

we at last obtain the transformed variance we originally needed:

Theorem 3.3.1. For $\mathcal{y} = My \otimes My$ and $V = var(y)$ as above, and $V_{\mathcal{y}} = var(\mathcal{y})$:

$$V_{\mathcal{y}} = (M \otimes M)[(V \otimes V)(I_{n^2} + I_{(n,n)}) + (Z \otimes Z)\Gamma(Z' \otimes Z')](M \otimes M).$$

Note that for data having $\gamma = 0$, for example normal data, we have simply:

$$V_{\mathcal{y}} = (MVM \otimes MVM)\ [I_{n^2} + I_{(n,n)}].$$

We conclude this section with an important application of our variance calculation.

We have seen in Chapter 1 that any translation invariant quadratic estimate has the form $h(y) = y'MAMy$ for some symmetric A. Hence such an estimator can also be expressed as a linear function of $\mathcal{y}$: $h(y) = g(\mathcal{y}) = t'\mathcal{y}$, for $t = vec\ A$. Here we now derive the covariance for any two such estimates, a result which will be of great utility when we derive our main optimality results later.

One form of the proof appears in Anderson's compendium-style thesis [1978], and this is essentially the line followed here, in that his argument concerns the variance of a single such estimator and this we extend. See also Rao [1971b].

We begin by setting r = vec A, t = vec B, so that

$$\begin{aligned} \mathrm{cov}[y'MAMy,\ y'MBMy] &= \mathrm{cov}[r'\mathcal{y},\ t'\mathcal{y}] \\ &= r'[\mathrm{var}(\mathcal{y})]t \\ &= r'[(M \otimes M)F(M \otimes M)]t \end{aligned}$$

$$= r'[(MVM \otimes MVM)(I_{n^2} + I_{(n,n)}) \ + \ (MZ \otimes MZ)\Gamma(Z'M \otimes Z'M)]t$$

$$= r'[(MVM \otimes MVM)(I_{n^2} + I_{(n,n)})]t \ + \ r'[(MZ \otimes MZ)\Gamma(Z'M \otimes Z'M)]t.$$

We now can apply [11] from 2.3 to conclude that

$$I_{(n,n)}t = I_{(n,n)}\ \mathrm{vec}\ A = \mathrm{vec}\ A = t,$$

so the covariance is

$$= r'[(MVM \otimes MVM)(I_{n^2}t + I_{(n,n)}t)] \ + \ r'[(MZ \otimes MZ)\Gamma(Z'M \otimes Z'M)]t$$

$$= r'[2(MVM \otimes MVM)]t \ + \ r'[(MZ \otimes MZ)\Gamma(Z'M \otimes Z'M)]t$$

Now check that

$$\begin{aligned} r'[2(MVM \otimes MVM)]t &= 2r'[\mathrm{vec}\ (MVM)t(MVM)] \\ &= 2(\mathrm{vec}\ B)'[\mathrm{vec}\ (MVM)A(MVM)] \\ &= 2\mathrm{tr}[B(MVM)A(MVM)] \\ &= 2\mathrm{tr}[(MBM)V(MAM)V], \end{aligned}$$

where we have used [9] from 2.3. Hence

$$\begin{aligned}\operatorname{cov}[y'MAMy,\ y'MBMy] &= 2\operatorname{tr}[B(MVM)A(MVM)] \\ &\quad + (\operatorname{vec} A)'(MZ \otimes MZ)\Gamma(MZ \otimes MZ)'(\operatorname{vec} B) \\ &= 2\operatorname{tr}[B(MVM)A(MVM)] \\ &\quad + [\operatorname{vec} Z'MAMZ]'\Gamma[\operatorname{vec} Z'MBMZ].\end{aligned}$$

This simplifies if we make the following observations: the elements of $v = \operatorname{vec} C$ which arise from diagonal elements of square C are, by inspection, those v_i with $i = 1 + j(m + 1)$, for $j = 0, 1, \ldots, m - 1$, One can also now check that these indices are exactly those diagonal elements of Γ which are possbily non-zero: the numbers $\gamma_i\sigma_i^4$, all other elements of Γ being zero. Further for $C = Z'MAMZ$ the partition of Z results in the diagonal elements of vec C being just the diagonal elements of $Z_i'MAMZ_i$ in turn, $i = 1, \ldots, k$. For now let's write $C_{[d]}$ for the vector which consists of the diagonal elements of C, so that we can finally conclude that

<u>Theorem 3.3.2.</u> For the translation invariant estimates $y'MAMy$ and $y'MBMy$:

$$\begin{aligned}\operatorname{cov}[y'MAMy,\ y'MBMy] &= 2\operatorname{tr}[A(MVM)B(MVM)] \\ &\quad + \sum \gamma_i\sigma_i^4 (Z_i'MAMZ_i)_{[d]}(Z_i'MBMZ_i)_{[d]}.\end{aligned}$$

Recalling that $y_* = My$ has $\operatorname{var}(y_*) = V_* = MVM$, and that any translation invariant estimator quadratic in y must be expressible in the form y_*Ay_* for some A, we also at last get

<u>Corollary 3.3.2.</u> For symmetric A and B and $y_* = My$:

$$\begin{aligned}\operatorname{cov}[y_*Ay_*,\ y_*By_*] &= 2\operatorname{tr}[AV_*BV_*] \\ &\quad + \sum \gamma_i\sigma_i^4 (Z_i'MAMZ_i)_{[d]}(Z_i'MBMZ_i)_{[d]}.\end{aligned}$$

3.4 The Second Linearization of the Basic Model.

It is possible to form a linear model for the components in a manner slightly different from that of [1]. In so doing the expression for var($\mathcal{y}$) simplifies, but weighing against this, constraints are introduced on the parameter vector to be estimated.

As before, start with vec [var(My)]:

$$\text{vec } [\text{var}(My)] = \text{vec } [E(My)(My)']$$

$$= \text{vec } M[\text{var}(y)]M$$

$$= \text{vec } MVM$$

$$= \text{vec } MZDZ'M.$$

Letting

$$\mathfrak{X} = MZ \otimes MZ,$$

we get

$$\text{vec } MZDZ'M = (MZ \otimes MZ)(\text{vec } D) = \mathfrak{X}\ (\text{vec } D).$$

Next let's put

$$\beta = \text{vec } D = \text{vec } (\oplus \sigma_i^2 I_{c_i}).$$

Then the new parameter vector β contains the components along with a great number of zeroes: β is constrained. Assume that Λ is a full rank matrix such that

$$\delta = \text{vec } B \text{ for some matrix } B = \oplus b_i I_{c_i} \quad \text{if and only if} \quad \Lambda\delta = 0.$$

With these definitions the linearization takes the form:

$$[1] \qquad E(\mathcal{y}) = \mathfrak{X}\beta, \text{ with } \beta \text{ such that } \Lambda\beta = 0.$$

Letting H be a matrix of maximal rank such that $H'\Lambda = 0$ implies $\beta =$

$H\sigma^2$, so that $\mathcal{X}\beta = \mathcal{X}_*\sigma^2$ for all $\sigma^2 \Rightarrow \mathcal{X}\beta = \mathcal{X}H\sigma^2 = \mathcal{X}_*\sigma^2 \Rightarrow \mathcal{X}_* = \mathcal{X}H$, thus connecting the two design matrices together.

Next, since $MVM = MZDZ'M$ we see also

$$\begin{aligned} MVM \otimes MVM &= MZDZ'M \otimes MZDZ'M \\ &= (MZ \otimes MZ)(D \otimes D)(Z'M \otimes Z'M) \\ &= \mathcal{X}(D \otimes D)\mathcal{X}'. \end{aligned}$$

Using [6] from 2.3

$$(I_{m^2} + I_{(m,m)})^2 = 2(I_{m^2} + I_{(m,m)}),$$

so letting

$$P = (1/\sqrt{2})(I_{m^2} + I_{(m,m)}),$$

and then checking that $P\Gamma = \Gamma P = P\Gamma P = \Gamma$, leads finally to

[2] $$\operatorname{var}(\boldsymbol{y}) = V_{\boldsymbol{y}} = \mathcal{X}P(D \otimes D + \Gamma)P\mathcal{X}'.$$

This last is the desired simplification, and as will be seen is suggestive of the results of Zyskind on optimal linear estimation, in that the variance for our model can be factored through the design matrix $\mathcal{X}$ of the model. It is this natural factorization, hinting at optimality, which is the principal usefulness of this form of the linearization.

We complete our introduction of this second linearization with a closer look at the contraints involved.

By construction we have a matrix Λ such that $\beta = \text{vec } D$ if and only if $\Lambda\beta = 0$. For a given Λ some of its rows will be found to be elements of the row space of $\mathcal{X}$, $= R(\mathcal{X})$, others not. This distinction will become important in our later study of optimal estimates, and then we will make use of the following recasting of the constraint matrix Λ:

Lemma 3.4.1. An equivalent formulation of the second linearization can be found using a constraint matrix Λ_* such that

$$\Lambda_* = \begin{bmatrix} \Lambda_1 \\ \Lambda_2 \end{bmatrix}$$

where $R(\Lambda_1) \cap R(\mathfrak{X}) = 0$, and $R(\Lambda_2) \subseteq R(\mathfrak{X})$.

Proof. Begin by forming a provisional Λ_2 which consists of all those rows of Λ contained in $R(\mathfrak{X})$. Work now by induction, considering the the first remaining row of Λ which is non-estimable and calling it the provisional Λ_1. At any later stage the next non-estimable row of Λ will be such that one of two things will occur: either the space obtained by adjoining it to our provisional Λ_1 is disjoint from $R(\mathfrak{X})$, in which case we add it to Λ_1, or there exists some element of Λ_1 such that it plus our test vector is in $R(\mathfrak{X})$. In this second case, discard our test vector and add the sum obtained to Λ_2. In this way new provisional Λ_1 and Λ_2 are formed, and after a finite number of steps a new matrix Λ_* is constructed as required. ∎

3.5 Additional Details of the Linearizations.

We conclude this chapter with a technical lemma which among other things simplifies even further our expression for Γ. We first introduce a set of structure matrices S_j and s_i. To do this let N[j] be the set of integers

$$N[j] = \{ \, (\sum_{i<j} c_i) + 1, \; \ldots, \; (\sum_{i<j} c_i) + c_j \, \}$$

Next use e_i = the ith unit vector in $\mathbf{R}_m$, and put

$$e_{(i)} = \text{diag } e_i \in \mathbf{R}_{m \times m}.$$

Then write

$$S_j = \sum_{i \varepsilon N[j]} e_{(i)} \otimes e_{(i)}, \quad \text{for } 1 \le j \le k,$$

$$s_i = MZe_{(i)}Z'M, \quad \text{for } 1 \le i \le m.$$

<u>Lemma 3.5.1.</u> [1] $\sum_{i \varepsilon N[j]} s_i = M_j.$

[2] $\mathfrak{X} S_j \mathfrak{X}' = \sum_{i \varepsilon N[j]} s_i \otimes s_i.$

[3] $\Gamma = \sum_{j=1} \gamma_j S_j (D \otimes D) = \sum_{j=1} \sigma_j^4 \gamma_j S_j.$

[4] $\mathfrak{X} \Gamma \mathfrak{X}'(\text{vec } a) = \text{vec } [\sum_{j=1} \sum_{i \varepsilon N[j]} \sigma_j^4 \gamma_j s_i (a) s_i],$

for $a \; \varepsilon \; \mathbb{R}_{n \times n}$.

<u>Proof.</u> [1] follows directly from the partitioning of Z; similarly for [2]. [3] is an exercise in keeping track of indices. [4] is a little more involved, and this we write out:

$$\mathfrak{X} \Gamma \mathfrak{X}'(\text{vec } a) = (MZ \otimes MZ)(\sum_{j=1} \gamma_j S_j)(MZ \otimes MZ)'(\text{vec } a)$$

$$= \sum_{j=1} \gamma_j (MZ \otimes MZ) \; S_j \; (MZ \otimes MZ)'(\text{vec } a)$$

$$= \sum_{j=1} \sum_{i \varepsilon N[j]} \sigma_j^4 \gamma_j (MZ \otimes MZ)(e_{(i)} \otimes e_{(i)})(MZ \otimes MZ)'(\text{vec } a)$$

$$= \sum_{j=1} \sum_{i \varepsilon N[j]} \sigma_j^4 \gamma_j (MZe_{(i)}Z'M \otimes MZe_{(i)}Z'M)(\text{vec } a)$$

$$= \sum_{j=1} \sum_{i \varepsilon N[j]} \sigma_j^4 \gamma_j \; \text{vec } [(MZe_{(i)}Z'M)a(MZe_{(i)}Z'M)]$$

$$= \sum_{j=1} \sum_{i \varepsilon N[j]} \sigma_j^4 \gamma_j \; \text{vec } [s_i (a) s_i]$$

$$= \text{vec}\left[\sum_{j=1}\ \sum_{i\varepsilon N[j]} \sigma_j^4 \gamma_j s_i(a) s_i\right]$$

and this completes the proof. $\blacksquare$

Chapter Four: The Ordinary Least Squares Estimates

4.1 Introduction.

With the linearization of the basic model and its covariance matrix at hand we now start on the estimation of the components. We'll do this by calculating the ordinary least squares estimates for our linear model, and by discussing what is meant by "estimable function" in our context. As will be shown the problem is that the covariance we've found is singular. This, along with the possibly sticky point that the reader may have already noticed: the covariance is a function of the parameters (components) we wish to estimate. These issues are sorted out here and the the ordinary least squares estimates are related to other variance component estimation methods.

4.2 The Ordinary Least Squares Estimates: Calculation.

Let's recall the linear form of the components problem:

$$E[\mathbf{y}] = \mathcal{X}_*\sigma^2, \quad \text{and} \quad var(\mathbf{y}) = V_{\mathbf{y}} = \mathcal{X}PQP\mathcal{X}',$$

with

$$\mathcal{X}_* = [\text{vec } M_1 \mid \text{vec } M_2 \mid \ldots \mid \text{vec } M]$$

$$\mathcal{X} = MZ \otimes MZ,$$

and

$$M_i = MZ_iZ_i'M, \ 1 \le i \le k, \quad \text{and} \quad M_k = M = I_n - \mathcal{P}_X.$$

Here and elsewhere, we will for convenience use $\mathcal{X}_*$ in the mean part, $\mathscr{X}$ in the covariance part of the model.

<u>Definition.</u> A vector estimate $\hat{\sigma}^2$ of $\sigma^2 = (\sigma_1^2, \sigma_2^2, \ldots, \sigma_k^2)$ is an <u>ordinary least squares</u> (<u>OLS</u>) estimate if it is a solution of the <u>normal equations</u>

$$\mathscr{X}_*'\mathscr{X}_*\hat{\sigma}^2 = \mathscr{X}_*'\mathscr{y}$$

Let's simplify these defining equations. Thus:

$$\mathscr{X}_*'\mathscr{X}_* = [\text{vec } M_1 \mid \text{vec } M_2 \mid \ldots \mid \text{vec } M]'$$

$$\times\ [\text{vec } M_1 \mid \text{vec } M_2 \mid \ldots \mid \text{vec } M]$$

$$= [(\text{vec } M_i)'(\text{vec } M_j)], \quad 1 \le i,j \le k.$$

That is, the matrix whose (i,j)th element is

$$(\text{vec } M_i)'(\text{vec } M_j) = \text{tr } M_iM_j,$$

where we use [9] of 2.3, so that

$$\mathscr{X}_*'\mathscr{X}_* = [(\text{tr } M_iM_j)].$$

Next

$$\mathscr{X}_*'\mathscr{y} = [\text{vec } M_1 \mid \text{vec } M_2 \mid \ldots \mid \text{vec } M]'\mathscr{y}$$

$$= [(\text{vec } M_1)'(My \otimes My) \mid (\text{vec } M_2)'(My \otimes My) \mid \ldots$$

$$\ldots \mid (\text{vec } M)'(My \otimes My)]'$$

$$= [\text{vec } (y'MM_1My) \mid \text{vec } (y'MM_2My) \mid \ldots \mid \text{vec } (y'MMMy)]'$$

$$= [(\text{vec } y'M_iy)], \quad \text{for } 1 \le i \le k.$$

Otherwise put, the OLS estimate $\hat{\sigma}^2$ solves the equations

$$[(\text{tr } M_iM_j)]\hat{\sigma}^2 = [(\text{vec } y'M_iy)],$$

or, letting A^- denote any g-inverse for a matrix A:

$$\hat{\sigma}^2 = [(\mathrm{tr}\ M_iM_j)]^-[(\mathrm{vec}\ y'M_iy)].$$

Furthermore, as can be shown in the standard linear model texts [e.g. Searle, 1971] the OLS estimate solves the problem

$$\min\ [(\mathcal{y} - \mathcal{X}_*\hat{\sigma}^2)'(\mathcal{y} - \mathcal{X}_*\hat{\sigma}^2)].$$

Similarly, one can formulate the constrained estimation problem as one of finding $\hat{\beta}$ such that

$$\min\ [(\mathcal{y} - \mathcal{X}\hat{\beta})'(\mathcal{y} - \mathcal{X}\hat{\beta})],\quad \text{with } \hat{\beta} \text{ such that } \Lambda\hat{\beta} = 0,$$

and then by standard linear model methods show that this version of the estimation problem has solution

$$\hat{\beta} = [I_{m^2} - G\Lambda_2'(\Lambda_2 G\Lambda_2')^{-1}\Lambda_2]\hat{\beta}_1,$$

where

$$\hat{\beta}_1 = G\mathcal{X}'\mathcal{y}\ ,$$

$$G = (\mathcal{X}'\mathcal{X} + \Lambda_1'\Lambda_1)^-$$

and Λ_1 and Λ_2 are the two halves of the constraint matrix as expressed in Lemma 3.4.1. As would be expected, for a suitable class of parametric functions of the components, the two estimation procedures lead to functions which are numerically identical as estimates of particular elements of the class, elements which we later show can be estimated in an optimal way. We now examine in detail this privileged class of parametric functions of the components, beginning with a closer look at the inner workings of our linear model for the components.

4.3 The Inner Structure of the Linearization.

With the linearized variance component model at hand and estimation our goal, we must now face some technicalities of the general linear model, issues which have the potential for seriously complicating our life in statistics. These difficulties are:

[1] The variance we calculated in Theorem 3.3.1 is in general singular:

$$V_{\mathcal{Y}} = \mathcal{X}PQP\mathcal{X}' = (M \otimes M)(Z \otimes Z)(PQP)(Z' \otimes Z')(M \otimes M)$$

with M generally not the identity and Z not of full rank, and with this holding regardless of the value of σ^2.

[2] The value of $V_{\mathcal{Y}}$ is itself in general a function of σ^2, the parameter vector we wish to estimate. That is, the variance is a function of the mean in the linear model.

[3] In general, with a singular variance, the data vector is restricted to a proper subspace of $\mathbf{R}_n$. To see this, suppose a given data vector z has variance $\mathcal{C}$ which is singular. For matrix A, let $A^{\perp}$ = a matrix of maximal rank such that $(A^{\perp})'A = 0$, and write $N = \mathcal{C}^{\perp}$. Then the following occurs:

$$\mathrm{var}(N'z) = N'\mathcal{C}N = 0 \;\Rightarrow\; N'z = \text{a constant, say } \delta$$

where the equation is valid off a set of probability zero, and the set in turn may be a function of α. Thus z is restricted to an affine subspace (a flat), and so is its expectation. Hence if the expectation is assumed to have linear structure, $E(z) = \mathcal{W}\alpha$ say, it follows that α will in general be subject to constraints:

$$N'z = \delta \;\Rightarrow\; E(N'z) = N'E(z) = N'\mathcal{W}\alpha = \delta.$$

In our situation note that since $\mathrm{var}(\mathcal{Y})$ is a function of σ^2 it follows $[\mathrm{var}(\mathcal{Y})]^{\perp}$ must also be presumed to be a function of σ^2. Thus we may further say that the constraining equation

$$[(\mathrm{var}(\mathcal{y}))^{\perp}]'\mathcal{X}_{*}\sigma^2 \;=\; \text{constant } \delta$$

is not necessarily linear in the components σ_i^2, and in fact the constant δ will itself be a function of σ^2.

Historically, [1] and [2] have created some problems, but the literature of estimation has shown that they have usually been surmountable. However, the issues raised in [3] have hardly been examined (see the review article by Kleefe [1977, p. 221]), much less resolved. For the variance component problem, these issues can be eluded as follows:

Since we assume that σ^2 can take on all real values, it is <u>unconstrained</u>, which makes the "constraining" equation, $N'\mathcal{X}_{*}\sigma^2 = \delta$, only a connecting relation between N, (unknown), $\mathcal{X}_{*}$ (known) and σ^2 (unknown), since N is in general a function of σ^2. Thus we may always argue for example that $a'\sigma^2 = b'\sigma^2$, for all σ^2, $\Leftarrow \Rightarrow$ $a = b$ for any $a, b \in \mathbb{R}_k$ but must still keep in mind that $N = (V_{\mathcal{y}})^{\perp}$ is in general a function of σ^2.

Also note that our argument for σ^2 is special to the kind of model we have written for the components, since more generally if var(z), or just N, is known not to be a function of α then $N'\mathcal{W}\alpha = \delta$ above is a linear constraint on α. The special nature of the linearization of the variance component model is further revealed if we assume that the data is restricted to the case:

$$\{\sigma_i^2 > 0,\ \gamma_i > -1,\ \text{for all } i\}.$$

This will hold if all the random components b_i have distributions which are continuous, unimodal and symmetric (see Kendall and Stuart [1977, p. 88]) -- for example, data with y normally distributed and no component degenerate. In this region of the parameter space N is functionally independent of σ^2. Still, we cannot conclude from this consequence of this specialization of (σ^2, γ) that σ^2 is constrained in the model we use for estimation: see the parallel discussion in Section 3.2.

4.4 Estimable Functions of the Components.

In light of the discussion above we are now free to make the standard

Definition. An estimable function of σ^2 is one of the form $q'\sigma^2$ for $q = (q_1, q_2, \ldots, q_k)'$ such that the function has a linear estimate $\ell'\mathcal{y}$ which is unbiased for $q'\sigma^2$.

Then

$$E[\ell'\mathcal{y}] = q'\sigma^2 \Leftarrow \Rightarrow \ell'E[\mathcal{y}] = \ell'\mathcal{X}_*\sigma^2 = q'\sigma^2 \text{ for all } \sigma^2$$

$$\Leftarrow \Rightarrow q' = \ell'\mathcal{X}_*,$$

so that

$$\ell'\mathcal{y} = y'Ay, \text{ for } A = \text{mat}_n\ell.$$

4.5 Further OLS Facts.

We complete this chapter by speaking of the connections between what we have called the OLS estimates and several other estimates. In fact our OLS estimates themselves made their first appearance in the literature under a much different guise, and we briefly discuss these interrelations.

To begin with, the OLS estimates are a specialization of the MINQUE estimates of Rao [1971a, 1971b] in that using a trivial weight matrix of all zeroes in the defining equations will generate the normal equations we have constructed above, from which the OLS estimates follow. Also, under this name MINQUE, the OLS estimates appear in many of the major statistical computer packages, such as

SAS®.

Next, the defining model for the OLS estimates, that is, the First or Second Linearization, can be solved in two other ways, and in so doing will generate the defining equations for two other standard estimation procedures for variance components. Thus, if generalized least squares is used to solve for the components in the First Linearization, then the Restricted Maximum Likelihood estimates result. And if the generalized least squares estimate is used first for the fixed-effects part of the original variance component model, and then generalized least squares is used in the dispersion-mean model based on untransformed data, then the usual Maximum Likelihood estimates result.

The complete details of these interconnections can be found in a monograph by S. R. Searle [1979].

Chapter Five: The Seely-Zyskind Results

5.1 Introduction.

We now take up the search for optimal unbiased estimates in certain quite general linear models, a search which historically led to one of the first truly general solutions to optimal estimation of variance components. This is accomplished through a careful presentation of results which were developed in a series of papers by Justus Seely and George Zyskind: Zyskind [1967], Seely [1971], and Seely and Zyskind [1971]. Then at the end of the chapter this material is applied to the original variance component problem.

Strictly speaking, these results are not required for us to obtain the general solution to the variance component estimation problem. They are, however, historically important, because they were the first of their kind, and because they stood for so long as one of the few general solutions to the problem. Therefore, their inclusion allows the procedures we present in later chapters to be put into context. We also comment on some technical rough spots that have appeared in the literature in the hope that a previously frustrating subject may be made merely difficult.

We also note that this practical and important material has, unfortunately, only recently began to find its way into the texts (for example Eaton [1984]) while its application to variance components has until now resided only in the journal literature.

Alternatively, one may view the material as the global solution to optimal unbiased estimation, and correspondingly, to the variance component estimation problem, in that conditions are found under which <u>every</u> estimable function has a best estimate. In the following chapter, we find the general solution to the component problem by looking for "local" conditions, conditions which are sufficient for a particular estimable function to have a best estimate.

5.2 The Generalized Gauss-Markov Theorem: Some History and Motivation.

The series of papers referred to above deal with extensions and applications of the Gauss-Markov Theorem on best unbiased linear estimation. This series represents only a small portion of the total research effort done along these lines: see also Rao [1967], Goldman and Zelen [1964], Kruskal [1968], Eaton [1970], and more recently Olsen, Seely, and Birkes [1976].

Here we present an extended version of the early Seely-Zyskind results. The detailed proof of our extension illustrates the caution required when working with such general linear models. The proof we give uses only the essential idea of the Seely-Zyskind results, which appears as our Lemma 5.4.1. This same idea is applied repeatedly in changing circumstances within the proof, making it somewhat harder to grasp intuitively, at least at first. However, the versatility of the Seely-Zyskind results is thereby made most visible.

The standard Gauss-Markov theorem runs as follows: In a linear model with covariance matrix of the form $\sigma^2 I_n$, every estimable function of the vector parameter has a best linear unbiased estimate, that is, an estimate of minimum variance in the class of all unbiased estimates which are linear in the data. Moreover, the theorem continues, this estimate is the ordinary least squares estimate, obtained from the normal equations for the model.

Much thought has gone into the problem of extending this basic and useful fact, for example, to covariance matrices not of the form constant times the identity. T. W. Anderson [1948] noticed that the same conclusions about bestness could be obtained for all models having covariance matrices which left the column space of the design matrix invariant. This important observation sat relatively unnoticed until 1965 or 1967 (the historical record is a little murky) when Zyskind began his more thorough study of this covariance condition and its implications (see Zyskind [1967]).

5.3 The General Gauss-Markov Theorem: Preliminaries.

The theorem we give is slightly more general than any presently available, but this generality is actually required, as will be seen, for the later discussion of of the Seely-Zyskind results.

We begin with specification of our most general linear model. Let y and e be random variables $\varepsilon\ \mathbb{R}_n$ such that:

$$y = X\beta + e,$$

with $E[e] = 0$,

X the given design matrix, $n \times p$

β a constant vector $\varepsilon\ \mathbb{R}_p$,

and $\operatorname{var}(y) = V$, assumed positive semidefinite, $n \times n$.

To these basic specifications we add:

[a] We allow β to be constrained, so that for some matrix Λ we have $\Lambda\beta = 0$. We assume Λ has full rank.

[b] We allow $V = \operatorname{var}(y)$ to be a function of β, written $V = V[\beta]$. We allow V to be singular and write N for a matrix of full rank such that $(N)'V = 0$, so that $N = V^{\perp}$, and where N is also possibly a function of β, so $N = N[\beta]$. We assume that V is p.s.d. for all allowable β.

[c] Other than that E(y) and var(y) should exist, we make no distributional assumptions about y and e.

As we have seen in Theorem 3.3.1, the linearization of the variance component model will necessarily involve using a covariance matrix which is itself a function of the parameters to be estimated, so that condition [b] above is critical to the study of the general variance estimation problem.

As usual write $\mathcal{R}(A)$ and $\mathcal{C}(A)$ for the row and column space of matrix A, and r(A) = rank(A). By BLUE we will mean an estimate which among all linear unbiased estimates of β, has smallest variance. We usually assume that the linear estimate $\ell'y$ has ℓ fixed and is not a function of β (while almost everything else in the analysis will be). Also, write A^- for any g-inverse of matrix A, and for $\mathcal{S}$ a subspace of $\mathbb{R}_n$, let the standard Euclidean orthogonal projection of $\mathbb{R}_n$ onto $\mathcal{S}$ be denoted by $\mathcal{P}_{\mathcal{S}}$.

Finally, denote the model described above with conditions [a], [b] and [c], by

$$\mathcal{MD}(y,\ X\beta,\ V;\ \Lambda).$$

In a model without constraints we drop the semicolon:

$$\mathcal{MD}(y,\ X\beta,\ V),$$

and in one having estimable and non-estimable constraints as in Lemma 3.4.1 we write

$$\mathcal{MD}(y,\ X\beta,\ V;\ \Lambda_1,\ \Lambda_2)$$

for Λ_1 non-estimable, Λ_2 estimable. The notation will thus reflect the presence of of any constraints imposed which are in addition to any derived from the possible singularity of V, as discussed in Section 4.3. Also, we will generally err on the side of redundancy, so that if y and Xβ, say, already imply that β is constrained by Λ, then we will sometimes still include Λ in the model specification.

At this point some old ground must be covered, since by allowing

V and hence N to be functions of the parameter β to be estimated, we do not have available a small body of work dealing with the most general form of an estimable function (Rao [1973b] and Harville [1981] for example). This work assumes that V and N are fixed, not a function of β. Unfortunately this assumption is not made explicit in these papers, nor in the earlier Zyskind [1967]. This omission leads to confusion, at least for for this reader, particularly when the results are applied to variance components. (see, for example, Seely and Zyskind [1971]).

To rescue from this tangle a useful notion of estimability we choose simply to use the following

Definition. An estimable function of β is one of the form $p'\beta$ for $p' \varepsilon \mathcal{R}(X)$.

It is not immediately clear how much is given up by this definition since the assumption of V and N being dependent on β disallows the techniques of Rao and Harville mentioned above. Thus, one cannot conclude that a parametric function $p'\beta$ having a linear unbiased estimate $\ell'y$ requires p' to be $\varepsilon\ \mathcal{R}(X)$ or even $p' = \ell'X + t'(N'X)$, for some $t \varepsilon \mathbb{R}_n$ (see Harville [1981]). This question we leave unanswered and will assume that the researcher finds all his $p'\beta$ of interest are of the form $p' \varepsilon \mathcal{R}(X)$.

Next, recall

Definition. In $\mathcal{MD}(y, X\beta, V; \Lambda)$ we say $\hat{\beta}$ is an ordinary least squares estimate (OLS) of β if it is of the form $\hat{\beta} = GX'y$ for $G = (X'X)^-$.

Let's observe first that the definition permits any choice of g-inverse to be made; however, standard linear model theory leads to the fact that $a'\hat{\beta}$ is invariant to the choice of g-inverse in $\hat{\beta} = GX'y$, given that $a'\beta$ is estimable. Second, note that the definition presumably ignores the presence of any constraints $\Lambda\beta = 0$. However, for non-estimable Λ_1, we can always choose to use G of the form

$$G = G_1 = (X'X + \Lambda_1'\Lambda_1)^-$$

and then again by standard methods show that $\hat{\beta}_1 = G_1X'y$ satisfies $\Lambda_1\hat{\beta}_1 = 0$. Moreover, letting Ω be the space of all allowable means E(y), then for any given system of constraints, matters can always be arranged so that if the constraints are all non-estimable, then the OLS we use is such that $\hat{\beta} = \mathcal{P}_{\Omega}(y)$. This fact does not hold up in the presence of estimable constraints, and this will be a key to the fuller understanding of the variance component problem later on.

5.4. The General Gauss-Markov Theorem: Statement and Proof.

We now come to the central lemma of the generalized Gauss-Markov results, at least in the form presented by Seely-Zyskind-Eaton-Rao. The result is Theorem 5 of Zyskind [1967].

Lemma 5.4.1. Suppose in $\mathcal{MD}(y, X\beta, V; \Lambda_1)$ with Λ_1 non-estimable constraints, that there is a matrix Q such that VX = XQ. Then there exists matrix a Q_* such that $V_*X_* = X_*Q_*$ in $\mathcal{MD}(y_*, X_*\beta, V_*; \Lambda_1)$, where

$$y_* = \begin{bmatrix} y \\ 0 \end{bmatrix} \qquad X_* = \begin{bmatrix} X \\ \Lambda_1 \end{bmatrix} \qquad V_* = \begin{bmatrix} V & 0 \\ 0 & 0 \end{bmatrix}$$

Proof. Let matrix Λ_1 be of full rank such that $\mathcal{R}(\Lambda_1) \cap \mathcal{R}(X) = 0$. Given an arbitrary constraint matrix Λ there is such a submatrix by Lemma 3.4.1. The dimension of $\mathcal{C}(X_*)$ is the row rank of X_* = row rank (X) + row rank (Λ_1) = column rank (X) + column rank (Λ_1) = column rank (X_+) + column rank(Λ_{1+}), where

$$(X_+)' = [\ X' \mid 0\ ] \quad \text{and} \quad (\Lambda_{1+})' = [\ 0 \mid (\Lambda_1)'\ ].$$

Letting $\mathcal{S}$ be the column space spanned by the columns of the two augmented matrices just above, then dim $\mathcal{S}$ = dim $\mathcal{C}(X_*)$, and since $\mathcal{S} \supseteq \mathcal{C}(X_*)$ we must have $\mathcal{S} = \mathcal{C}(X_*)$. (Note the confusing statement at the second line from the bottom on p. 1104 of Zyskind [1967]; the line should read "... and the column spaces of X_+ and Λ_{1+} are disjoint".)

Thus the column bases of X and Λ_1 can be used to form a basis for X_* through the addition of trailing and leading components, respectively, which are zero: if x_i, i = 1, ..., r(X), and z_j, j = 1, ..., r(Λ_1), are a basis for X and Λ_1 respectively, then

$$x_{i+} = \begin{bmatrix} x_i \\ 0 \end{bmatrix} \quad \text{and} \quad z_{j+} = \begin{bmatrix} 0 \\ z_j \end{bmatrix} \quad \text{all } i, j$$

are all n-vectors which together form a basis for X_*. Next, it is straightforward linear algebra to show that VX = XQ if and only if X has a basis for its column space which consists of eigenvectors of V. Since we have seen that the set

$$\{x_{i+}, z_{j+} \mid \text{all } i, j\}$$

is a basis for $C(X_*)$ which is composed of eigenvectors of V_*, using the linear algebra fact just mentioned completes the proof. ∎

Let's note first that the lemma is dependent on Λ_1 being non-estimable; it does not hold for Λ_1 having any estimable rows. Thus, non-estimable constraints can always be imposed without penalty to the condition VX = XQ. Second, observe that from the lemma we also find that <u>if VX = XQ, for Λ_1 non-estimable, then there exists a matrix Q_* such that $\Lambda_1 Q_* = 0$, as well as $VX = XQ_*$.</u>

<u>Lemma 5.4.2.</u> In $\mathcal{MD}$(y, Xβ, V) if VX = XQ then for any U such that $U'U = I_n$ then putting $V_1 = UVU'$ and $X_1 = UX$ gives $V_1X_1 = X_1Q$.

Here, note that V_1 = var(Uy) so that an orthogonal transformation may also be imposed without cost to the condition $VX = XQ$.

Lemma 5.4.3. In $\mathcal{MD}(y, X\beta, V)$ suppose that $VX = XQ$ and U is a matrix such that

[1] $\mathcal{C}(U') \subseteq \mathcal{C}(X)$,

[2] $v'Uy$ is unbiased for $p'\beta$ for vector $v \in \mathbb{R}_n$,

[3] β is not constrained by construction or by the possible singularity of V.

Then $v'Uy$ is BLUE for $p'\beta$.

Proof. Fix β. Since $\mathcal{R}(U) \subseteq \mathcal{R}(X')$ we can find b such that $b'X' = v'U$, so that $b'X'y$ is linear and unbiased for $p'\beta$.

Suppose now that $k'y$ is any other linear unbiased estimate for estimable $p'\beta$, so that $p' = k'X$, and also $p' = b'X'X$, since β is unconstrained. Then

[a] $\text{var}(v'Uy) = b'X'VXb = b'X'XQb = p'Qb$;

[b] $\text{var}(k'y) = k'Vk$;

[c] $\text{cov}(v'Uy, k'y) = b'X'Vk = b'QX'k = b'Qp = p'Qb$.

Thus

$$\begin{aligned}\text{var}(v'Uy - k'y) &= p'Qb + k'Vk - 2p'Qb \\ &= k'Vk - p'Qb \\ &= \text{var}(k'y) - \text{var}(v'Uy).\end{aligned}$$

But necessarily $\text{var}(v'Uy - k'y) \geq 0$ so that $\text{var}(k'y) \geq \text{var}(v'Uy)$. ∎

Lemma 5.4.4. In $\mathcal{MD}(y, X\beta, V; \Lambda_1)$ with $\Lambda_1\beta = 0$ non-estimable constraints, where p, β and U are as in Lemma 5.4.3, if $VX = XQ$ then $v'Uy$ is BLUE for $p'\beta$.

<u>Proof.</u> We use y_*, X_* and V_* as in Lemma 5.4.1 and will refer to a note following the proof of that lemma. Then for U_+ such that $(U_+)' = [\, U' \mid 0 \,]$, one gets $\mathscr{C}(U_+) \subseteq \mathscr{C}(X_*)$, and $v'U_+y_* = v'Uy$.

Now start with $k_+y_* = k'y$, for $k_+ = [k', 0]$ and write

$$k_+y_* = k_+y_* - v'U_+y_* + v'U_+y_*,$$

where it is assumed that $k'y$ is unbiased for $p'\beta$. Then

$$\operatorname{var}(k_+y_*) = \operatorname{var}(k_+y_* - v'U_+y_*) + \operatorname{var}(v'U_+y_*)$$
$$+ 2\operatorname{cov}(k_+y_* - v'U_+y_*, v'U_+y_*)$$

Since

$$E[k_+y_* - v'U_+y_*] = (k_+ - v'U_+)E[y_*] = 0,$$

we see that $(k_+ - v'U_+)$ is orthogonal to the space spanned by all means $E(y_*)$, $= \Omega_*$, say. Then using the note following Lemma 5.4.1 we know that there is matrix Q_* such that $\Lambda_1Q_* = 0$, while $V_*X_* = X_*Q_*$, hence

$$\operatorname{cov}(k_+y_* - v'U_+y_*, v'U_+y_*) = (k_+ - v'U_+)V_*(U_+v),$$

and

$$V_*(U_+v) = V_*(X_*H_*)v = (X_*Q_*)H_*v,$$

for some matrix H_*, since $\mathscr{C}(U') \subseteq \mathscr{C}(X)$, and $V_*X_* = X_*Q_*$.

Also,

$$(\Lambda_1Q_*)H_*v = 0,$$

as $\Lambda_1Q_* = 0$, so finally $X_*(Q_*H_*v) \subseteq \Omega_*$, and consequently the covariance is zero. Hence necessarily $\operatorname{var}(k'y) = \operatorname{var}(k_+y_*) \geq \operatorname{var}(v'U_+y_*) = \operatorname{var}(v'Uy)$. ∎

At this point let's note, as earlier, that by using $U = GX'$ with

$$G = G_1 = (X'X + \Lambda_1'\Lambda_1)^-,$$

and by putting $\hat{\beta}_1 = G_1X'y$, we get $v'Uy = p'\hat{\beta}_1$ as an OLS estimate which is BLUE for $p'\beta$ and which satisfies $\Lambda_1\hat{\beta}_1 = 0$. Also, the key issue with constraints $\Lambda\beta = 0$, given that $VX = XQ$, is that to obtain optimality one only needs to add $\Lambda Q = 0$. This holds no matter what the constraints might be, estimable, non-estimable or some mixture, as can be seen from a careful examination of the proof of Lemma 5.4.4. Although we won't use this fact in our later work, we set it out as a result in itself:

<u>Lemma 5.4.5</u>. In $\mathcal{MD}(y, X\beta, V; \Lambda_1, \Lambda_2)$ with p, β and U as in Lemma 5.4.3, if $VX = XQ$,; with $\Lambda Q = 0$, the $v'Uy$ is BLUE for $p'\beta$.

Finally observe that the proofs of Lemmas 5.4.3, 5.4.4, and 5.4.5 each work at <u>any fixed β</u>, so that they hold for all V of the form $V = V[\beta]$. Capitalizing on this important fact let's set forth the following

<u>Definition.</u> We call an estimate $k'y$ <u>Gauss-Markov</u> if

[1] $k'y$ is unbiased for $p'\beta$;

[2] k is possibly a function of β and of V: $k = k[\beta, V]$;

[3] at each β and V, $k'y$ has smallest variance among all estimates satisfying conditions [1] and [2].

Our notion of Gauss-Markov is a consistent generalization of that which appears in Kruskal [1968] and Eaton [1970]. It is also related to a generalization of the allowable estimating functions as studied by Rao [1973a] and Harville [1981]. Rao considers functions essentially of the form $k'y$ for $k = k[N'y]$, and looks at $k = k[d]$ for $\delta = N'X\beta$. But as we have seen $N'y = \delta$ (a.e.) so these two notions coincide up to a set of measure zero, which may itself depend on β. Our definition grants the possibility of k being any function of β and V. Alternatively, using the idea of "locally best" as studied by LaMotte [1976], we might also say that Gauss-Markov for us means "formally locally best" for situations where the covariance is a dependent function of β. Finally, these Gauss-Markov estimates will have, for the time being, a largely formal character, the point being that we later show how the invariance condition VX = XQ will lead to equality of the Gauss-Markov estimates and the OLS estimates, thus imparting a quite general optimality to the OLS values.

Given our new definition and Lemma 5.4.4 above we can now state:

Theorem 5.4.6. If VX = XQ in $\mathcal{MD}(y, X\beta, V; \Lambda_1)$ with non-estimable constraints $\Lambda_1\beta = 0$ and β otherwise unconstrained, then the OLS estimates are Gauss-Markov.

We turn now to the study of the converse of this Theorem, and do this by reformulating the proofs of Theorem 1 and 2 in Kruskal [1968]; it is from this work that our notion of Gauss-Markov originates.

To begin, let's recall that since V is symmetric and assumed p.s.d, we have available an orthogonal decomposition of $\mathbb{R}_n$, with $\mathbb{R}_n = \mathcal{R} \oplus \mathcal{N}$, where $\mathcal{R}$ is the range space of V, $\mathcal{N}$ is the null space of V, and where $\mathcal{N} = \mathcal{C}(V^{\perp})$. Note that V is an isomorphism of $\mathcal{R}$ so that there is a linear transformation W such that WV = VW = I on $\mathcal{R}$. Extending W to all of $\mathbb{R}_n$ by putting $W(\mathcal{N}) = 0$ leads to W well-defined on $\mathbb{R}_n$. Similarly we define $W^{\frac{1}{2}}$ and $W^{-\frac{1}{2}}$ so that $W^{\frac{1}{2}}W^{\frac{1}{2}} = W$, and $W^{-\frac{1}{2}}W^{\frac{1}{2}} = I$ on $\mathcal{R}$, = 0 on $\mathcal{N}$.

Suppose now that $E(y) = X\beta = \mu_y$, and that the space of all possible μ_y is denoted by Ω. One then checks that with probability one, $y \in \mathcal{R} + \mathcal{P}_{\mathcal{N}}(\mu_y)$, for $\mathcal{P}_{\mathcal{N}}$ the for $\mathcal{P}_{\mathcal{N}}$ the projection of $\mathbb{R}_n$ onto $\mathcal{N}$. Note that the set of measure zero in question may be a function of β.

Next letting v be any vector $\varepsilon\ \Omega$ such that $\mathscr{P}_{\mathscr{N}}(v) = \mathscr{P}_{\mathscr{N}}(\mu_y)$, we also get $\mathscr{P}_{\mathscr{N}}(v) = \mathscr{P}_{\mathscr{N}}(y)$, with probability one. Now let $z = y - v$ so that with $\mathscr{P}_{\mathscr{N}}(z) = 0$ with probability one, and $\text{var}(z) = V$, with $E(z) = \mu_z$ say, spanning $\Omega \cap \mathscr{R}$. Finally, let $t = W^{\frac{1}{2}}(z)$ so that $E(t) = \mu_t$ say, $= W^{\frac{1}{2}}(\mu_z) = W^{\frac{1}{2}}(\mu_y - v)$, and $\text{var}(t) =$ the identity on $\mathscr{R}$, and is zero on $\mathscr{N}$. Thus $\text{var}(t) = \mathscr{P}_{\mathscr{R}}$.

Set $\Omega_* = W^{\frac{1}{2}}(\Omega \cap \mathscr{R})$ so μ_t spans Ω_*, and then form the r.v.

$$\hat{\mu} = \hat{\mu}(t) = a'W^{\frac{1}{2}}\mathscr{P}_{\Omega_*}(t)$$

for a given vector a. As one finds, this provides an unbiased estimate of $a'\mu_t$. Next write

$$b't = (b - c)'t + c't$$

for a given vector b, getting then

$$\begin{aligned}\text{var}(b't) &= \text{var}[(b - c)'t] + \text{var}(c't) + 2\text{cov}[(b - c)'t,\ c't] \\ &= (b - c)'\mathscr{P}_{\mathscr{R}}(b - c) + c'\mathscr{P}_{\mathscr{R}}(c) + 2(b - c)'\mathscr{P}_{\mathscr{R}}(c).\end{aligned}$$

Note that if $E[(b - c)'t] = 0$ then also $(b - c)'\mu_t = 0$ so $(b - c)\ \varepsilon\ (\Omega_*)^{\perp}$. Further, if $c\ \varepsilon\ \Omega_*$ then $c = W^{\frac{1}{2}}(d)$ for some $d\ \varepsilon\ \Omega \cap \mathscr{R}$, while $W^{\frac{1}{2}}$ is onto $\mathscr{R}$, so $c\ \varepsilon\ \mathscr{R}$, and necessarily $\mathscr{P}_{\mathscr{R}}(c) = c$. For $c' = a'W^{\frac{1}{2}}\mathscr{P}_{\Omega_*} = (\mathscr{P}_{\Omega_*}W^{\frac{1}{2}}(a))'$ we get $c\ \varepsilon\ \Omega_*$, so that

$$2(b - c)'\mathscr{P}_{\mathscr{R}}(c) = 2(b - c)'c = 0,$$

and

$$\text{var}(b't) \geq \text{var}(c't).$$

It now follows that at each fixed β, $\hat{\mu}(t)$ has smaller variance than any other linear unbiased estimate $b't$ of $a'\mu_t$, where we can allow b to be a function of β. We have thus shown that $\hat{\mu}(t)$ is a Gauss-Markov estimate for the parametric function $a'\mu_t$.

Next, we have $\mathcal{P}_{\mathcal{N}}(y) = \mathcal{P}_{\mathcal{N}}(z + v) = \mathcal{P}_{\mathcal{N}}(z) + \mathcal{P}_{\mathcal{N}}(v) = 0$ with probability one, but $\mathcal{P}_{\mathcal{N}}(z) = 0$ with probability one, so $v \in \mathcal{R}$, in fact $v \in \Omega \cap \mathcal{R}$, recalling that $v \in \Omega$. Hence

$$\begin{aligned} a'\hat{\mu}(t) &= a'W^{\frac{1}{2}}\mathcal{P}_{\Omega_*}(t) \\ &= a'W^{\frac{1}{2}}\mathcal{P}_{\Omega_*}[W^{-\frac{1}{2}}(z)] \\ &= a'W^{\frac{1}{2}}\mathcal{P}_{\Omega_*}W^{-\frac{1}{2}}(y - v) \\ &= a'W^{\frac{1}{2}}\mathcal{P}_{\Omega_*}W^{-\frac{1}{2}}(y) \quad - \quad a'v. \end{aligned}$$

Recalling that the vector v is given and fixed, not a r.v., we have thus found that $\hat{\mu}(t)$ leads to an estimate based on y which is Gauss-Markov for the parametric function $a'\mu_y$.

Letting $\hat{\mu}(z) = W^{\frac{1}{2}}\mathcal{P}_{\Omega_*}(z)$ serves to define

$$a'\hat{\mu}(y) = a'\hat{\mu}(z) + a'v = a'W^{\frac{1}{2}}\mathcal{P}_{\Omega_*}W^{-\frac{1}{2}}(y).$$

Considering our problem of finding a converse for Theorem 5, let's assume now that the OLS for $a'\mu_y$ is Gauss-Markov. As the constraints are non-estimable we may use our earlier discussion to select the OLS $\hat{\beta}$ to be $\hat{\beta} = \mathcal{P}_{\Omega}(y)$. Then we must have

$$a'\mathcal{P}_{\Omega}(y) = a'\hat{\mu}(y), \text{ for all } a.$$

But then

$$\mathcal{P}_{\Omega}(y) = \hat{\mu}(y).$$

Defining $\mathcal{P}^* = W^{\frac{1}{2}}\mathcal{P}_{\Omega_*}W^{-\frac{1}{2}}$ we can check that $\mathcal{P}^*$ is the orthogonal projection onto Ω_* relative to the p.s.d. inner product $[a, b] = a'Wb$. Hence the condition above is equivalent to

$$\mathcal{P}^*(y) = \mathcal{P}_{\Omega}(y),$$

but note that we cannot go on to conclude $\mathcal{P}^* = \mathcal{P}_{\Omega}$ on all of $\mathbb{R}_n$, since for singular V the r.v. y need not assume all possible $v \in \mathbb{R}_n$, as for example $N'y = d$ with probability one. This last step apparently is erroneously taken in Kruskal [1968, Theorem 1]. We also point out that in Theorem 2 of Kruskal [1968] it is apparently asserted that relative to the p.d. inner product $[a, b] = a'V^{-\frac{1}{2}}b$, for non-singular V, the adjoint of any matrix A is equal to its transpose A', so that $[a, Ab] = [A'a, b]$. However, the correct adjoint for matrix A relative to this product is $V^{\frac{1}{2}}A'V^{-\frac{1}{2}}$. Still, the proof of his Theorem 2 goes through when the proper adjoint is used.

Continuing with our proof of the converse, consider the covariance

$$\text{cov}[x'\mathcal{P}^*(y), w'(I_n - \mathcal{P}^*)(y)] = x'(\mathcal{P}^*)V(I_n - \mathcal{P}^*)'w,$$

for all $x, w \in \mathbb{R}_n$,

$$= x'W^{\frac{1}{2}}\mathcal{P}_{\Omega_*}W^{-\frac{1}{2}}(I_n - W^{-\frac{1}{2}}\mathcal{P}_{\Omega_*}W^{\frac{1}{2}})w$$

$$= x'W^{\frac{1}{2}}\mathcal{P}_{\Omega_*}W^{\frac{1}{2}}w - x'W^{\frac{1}{2}}\mathcal{P}_{\Omega_*}W^{\frac{1}{2}}w = 0.$$

Consequently

$$\text{cov}[x'\mathcal{P}_\Omega(y), w(I_n - \mathcal{P}_\Omega)(y)] = 0$$

so that

$$x'\mathcal{P}_\Omega V(I_n - \mathcal{P}_\Omega)'w = 0,$$

for all $x, w \in \mathbb{R}_n$, so finally

$$\mathcal{P}_\Omega V(I_n - \mathcal{P}_\Omega) = 0.$$

One checks that this last holds if and only if $V\Omega \subseteq \Omega$, and that this occurs if and only if $VX = XQ$ for some matrix Q.

Finally, standard linear model theory tells us that any OLS in a model without non-estimable constraints remains OLS (a solution to the normal equations) in the model with them, and that the class of estimable functions does not change in the presence of such constraints, even though a given representation of $a'\mu_y$ might. Pulling these comments together with our work above we can now state the desired converse to Theorem 5.4.6 as:

Theorem 5.4.7. If, in $\mathcal{MD}(y, X\beta, V; \Lambda_1)$ with $\Lambda_1\beta$ non-estimable and β otherwise unconstrained, the OLS estimates are Gauss-Markov for all estimable functions $a'\mu_y$, then $VX = XQ$ for some matrix Q.

5.5. The Zyskind Version of the Gauss-Markov Theorem.

With Theorems 5.4.6 and 5.4.7 in hand we take the next step towards optimal variance component estimation by deriving an extension of an important result of Zyskind [1967]. This is, in light of the above, just a restatement of these main results in the setting of constrained estimation, so in one sense nothing fundamentally new is accomplished. However, by formulating the optimal estimation problem in terms of the constrained Second Linearization discussed in Section 3.4, we are able to describe the obstruction to obtaining an optimal estimate for every estimable function as one tied up with a set of estimable constraints. This, we believe, adds insight to our understanding of the overall estimation problem.

The result we now prove is an extension of Zyskind [1967] in that we explicitly allow V to be a function of β, and this is important since it is basic in the context of variance component estimation. The complete specification of our model is that given in Section 5.4.

Definition. For matrices B and C we say B is C-invariant if $CB \subseteq \mathcal{C}(B)$; equivalently

$$\mathcal{P}_B C \mathcal{Q}_B = 0.$$

Before beginning let's note that for $\Lambda_2\beta = 0$, estimable constraints, we can find a matrix B such that $\Lambda_2 = A'X$, and $\mathscr{C}(A) \subseteq \mathscr{C}(X)$, since $\mathscr{C}(X) = \mathscr{C}(XX')$. Hence for a matrix C we will for brevity say that Λ_2 is C-invariant when A is C-invariant.

Theorem 5.5.1. Given $\mathscr{MD}(y, X\beta, V, \Lambda_1, \Lambda_2)$ with $\Lambda_1\beta = 0$ non-estimable constraints, $\Lambda_2\beta = 0$ estimable constraints, and β otherwise unconstrained, suppose that both X and Λ_2 are V-invariant. Then $p'\hat{\beta}$ is Gauss-Markov for estimable $p'\beta$, where

$$\hat{\beta} = [I_n - G\Lambda_2(\Lambda_2 G\Lambda_2)^{-}\Lambda_2]\hat{\beta}_1,$$

for $\hat{\beta}_1 = GX'y$, and $G = (X'X + \Lambda_1\Lambda_1)^{-}$.

Proof. We begin with $VX = XQ$ for some matrix Q, and assume non-estimable con- straints $\Lambda_1\beta = 0$ are in force. Then by Lemma 5.4.1 there is matrix Q_1 such that for

$$y_1 = \begin{bmatrix} y \\ 0 \end{bmatrix} \qquad X_1 = \begin{bmatrix} X \\ \Lambda_1 \end{bmatrix} \qquad V_1 = \begin{bmatrix} V & 0 \\ 0 & 0 \end{bmatrix}$$

we have $V_1X_1 = X_1Q_1$, that is, X_1 is V_1-invariant.

Next, for $\Lambda_2\beta = 0$ estimable constraints, we know there is matrix A with $\Lambda_2 = A'X$, and $\mathscr{C}(A) \subseteq \mathscr{C}(X)$, so we can use the proof of Lemma 5.4.1 to see that A_1 such $(A_1)' = [\ A' \mid 0\ ]$, has $\mathscr{C}(A_1) \subseteq \mathscr{C}(X_1)$. Also, we can see that Λ_2 is V_1-invariant.

Letting $\mathscr{P}_1$ be the projection onto $\mathscr{C}(A_1)$, and $\mathscr{Q}_1 = I - \mathscr{P}_1$, consider the decomposition $y_1 = \mathscr{P}_1(y_1) + \mathscr{Q}_1(y_1)$. By V_1-invariance we have $\mathscr{P}_1V_1\mathscr{Q}_1 = 0$ so that $\mathrm{var}(k'y) \geq \mathrm{var}[k'\mathscr{Q}_1(y_1)]$ for any constant k, and since $E[\mathscr{P}_1(y_1)] = \mathscr{P}_1(X_1\beta) = A_1\beta = 0$, it follows that we need only consider linear estimates based on $\mathscr{Q}_1(y_1)$ in our search for minimum variance unbiased estimates.

Hence consider $y_2 = \mathscr{Q}_1(y_1)$, $X_2 = \mathscr{Q}_1(X_1)$, $V_2 = \mathscr{Q}_1V_1\mathscr{Q}_1$. The proof will be complete if we can verify the following five facts:

(1) with respect to X_2 the constraints $\Lambda_2\beta = 0$ are now <u>non-estimable</u>;

(2) with respect to X_2 the constraints $\Lambda_1\beta = 0$ remain non-estimable;

(3) X_2 is V_2-invariant;

(4) the class of functions estimable with respect to X_2 is exactly the same as the class estimable with respect to X;

(5) the OLS based on y_2 and X_2 is $\hat{\beta}$ as given in the statement of the Theorem.

We work on proving these in turn.

(1) Suppose to the contrary that there exists vectors a and b such that $a'A'X = b'X_2$. Then

$$a'A_1X - b'\mathcal{Q}_1X = 0 \;\Rightarrow\; (a'A_1 - b' + b'\mathcal{P}_1)X = 0$$

$$\Rightarrow\; a'A_1A_1 - b'A_1 + b'\mathcal{P}_1A_1 = 0,$$

since $\mathcal{C}(A_1) \subseteq \mathcal{C}(X_1)$. As $\mathcal{P}_1A_1 = A_1$, this means

$$a'A_1A_1 = 0 \;\Rightarrow\; a'A_1A_1a = 0 \;\Rightarrow\; A_1a = 0.$$

But we can choose A_1 of full column rank, so that only $a = 0$ is possible.

(2) For this, simply observe that

$$X_2 = \mathcal{Q}_1X_1 = [\;(\mathcal{Q}_AX)' \mid 0\;]'$$

for $\mathcal{Q}_A = I - \mathcal{P}_A$, so $\mathcal{R}(\mathcal{Q}_1X_1) \subseteq \mathcal{R}(X)$. Then

$$\mathcal{R}(\Lambda_1) \cap \mathcal{R}(X) = 0 \;\Rightarrow\; \mathcal{R}(\Lambda_1) \cap \mathcal{R}(X_2) = 0.$$

(3) Let's start by noting that $\mathscr{C}(A_1) \subseteq \mathscr{C}(X_1)$ implies we can write X_1 as

$$X_1 = (U_1 \mid U_2)H,$$

for matrices U_1, U_2 and H such that $\mathscr{C}(U_1) = \mathscr{C}(A_1)$, and $\mathscr{C}(U_2) = \mathscr{C}[(A_1)^{\perp}] \cap \mathscr{C}(X_1)$.

Then

$$V_2X_2 = (\mathscr{Q}_1V_1\mathscr{Q}_1)(\mathscr{Q}_1X_1) = \mathscr{Q}_1V_1\mathscr{Q}_1X_1$$

since $(\mathscr{Q}_1)^2 = \mathscr{Q}_1$,

$$= (I - \mathscr{P}_1)V_1(I - \mathscr{P}_1)X_1$$

$$= V_1X_1 - \mathscr{P}_1V_1X_1 - V_1\mathscr{P}_1X_1 + \mathscr{P}_1V_1\mathscr{P}_1X_1.$$

By construction now

$$\mathscr{P}_1X_1 = \mathscr{P}_1(U_1 \mid U_2)H$$

$$= [\mathscr{P}_1U_1 \mid \mathscr{P}_1U_2]H$$

$$= (U_1 \mid 0)H$$

so that $\mathscr{P}_1X_1 = X_1T$ for some matrix T.

Next, since $V_1X_1 = X_1Q_1$ we see

$$V_2X_2 = X_1Q_1 - \mathscr{P}_1(X_1Q_1) - V_1(X_1T) + \mathscr{P}_1V_1(X_1T)$$

$$= X_1Q_1 - X_1TQ_1 - X_1Q_1T + X_1TQ_1T = X_1Q_2,$$

for some matrix Q_2. But then

$$V_2X_2 = \mathscr{Q}_1V_1\mathscr{Q}_1X_1 = \mathscr{Q}_1(\mathscr{Q}_1V_1\mathscr{Q}_1X_1) = \mathscr{Q}_1V_2X_2 = \mathscr{Q}_1X_1Q_2 = X_2Q_2,$$

and the V_2-invariance of X_2 is demonstrated.

(4) Let $p' = k'X$. Then for k_1 such that $(k_1)' = (k' \mid 0)$, we get

$p' = (k_1)'X_1$. Also, $\mathscr{P}_1 = A_1(A_1A_1)^-A_1$, so that $\mathscr{P}_1X_1\beta = A_1(A_1A_1)^-A_1X_1\beta$ $= A_1(A_1A_1)^-\Lambda_2\beta = 0$, so that $p'b = (k_1)'X_1\beta = (k_1)'X_1\beta - (k_1)'\mathscr{P}_1X_1\beta =$ $(k_1)'X_2\beta = (p_1)'\beta$ for $(p_1)' = (k_1)'X_2$.

(5) It remains now to calculate the estimate which is the OLS of $p'\beta$ with respect to the design matrix X_2 and based on the transformed data vector $y_2 = \mathscr{Q}_1y_1$. We set this derivation out in several steps and leave the details as an exercise for the reader. Thus the estimate of the Theorem can be found as shown by verifying that

(a) for G any g-inverse of $X'X$, it follows that GX' and $G'X'$ are g-inverses of X;

(b) given $A'X = \Lambda_2$ it follows that $A' = \Lambda_2GX' = \Lambda_2G'X'$, and $A'A = \Lambda_2G\Lambda_2$;

(c) for $G = (X'X + \Lambda_1\Lambda_1)^-$ it follows that $\Lambda_1G\Lambda_2 = 0$ (Hint: let $B = (X'X + \Lambda_1\Lambda_1)^-$ and begin with $BGB = B$; use $\mathscr{R}(\Lambda_1\Lambda_1) \cap \mathscr{R}(X'X) = 0$ and conclude that $\Lambda_1GX' = 0$, so $\Lambda_1G(X'A) = \Lambda_1G\Lambda_2 = 0$).

Using (a) - (c) shows that $\hat{\beta}$ as stated above is the desired OLS and that $\hat{\beta}$ is such that $\Lambda_1\hat{\beta} = 0$ and $\Lambda_2\hat{\beta} = 0$, and this concludes the proof. ∎

5.6 The Seely Condition for Optimal Unbiased Estimation.

In this concluding section we apply our results on Gauss-Markov estimation to our original variance component problem. Let's begin with the

Definition. Let $Sp(\sigma^2)$ be the real linear space in $Sym(n)$ spanned by all the $M_i = MZ_iZ_i'M$. Let $Sp(\sigma^2)^\perp$ be the trace product orthocomplement of $Sp(\sigma^2)$ in $Sym(n)$, that is all $a \; \varepsilon$ $Sym(n)$ such that $tr(aM_i) = (a, M_i) = 0$ for all i.

In 1971 Seely introduced the following idea, stated in our notation:

<u>Definition.</u> $Sp(\sigma^2)$ is said to be a <u>quadratic subspace</u> if $a^2 \; \varepsilon \; Sp(\sigma^2)$ for all $a \; \varepsilon \; Sp(\sigma^2)$. Equivalently, aba $\varepsilon \; Sp(\sigma^2)$ for all a, b ε $Sp(\sigma^2)$.

Then he proved the following:

<u>Theorem 5.6.1 (Seely [1971])</u> Suppose that $y \; \varepsilon \; \mathbf{R}_n$ is distributed as a normal variable with mean zero and variance $V = \Sigma \; \sigma_i^2 V_i$. Then every estimable function of the components has a uniformly minimum variance, quadratic, unbiased estimate if and only if $Sp(\sigma^2)$ is a quadratic subspace.

We will prove a slightly more general version of this result. (Note though that if our data y does not have mean zero then we consider the reduced model for $y_* = My$ with $M = I - \mathscr{P}_{\mathscr{X}}$, and $var(y_*) =$ $V_* = \Sigma \; \sigma_i^2 M_i$.) We'll begin with some alternative formulations of the quadratic subspace condition. In a later chapter we show how the condition is actually that $Sp(\sigma^2)$ is a <u>Jordan algebra</u>, and in this context we also note that "quadratic" has historically had a distinctly different meaning: see Section 7.7.

Recall that $\mathscr{y} = My \otimes My$. Then from Chapter Three:

$E(\mathscr{y}) = E(My \otimes My)$

$\qquad = vec(MZDZ'M) = (MZ \otimes MZ)vec(D) = \mathscr{X}vec(D) = \mathscr{X}\beta; \quad \Lambda\beta = 0,$

using the assumed normality of the data (so $\gamma = 0$) and the Second

Linearization.

With the First Linearization we have simply

$$E(\mathcal{y}) = \mathcal{X}_*\sigma^2; \quad \sigma^2 \text{ unconstrained.}$$

Also

$$\text{var}(\mathcal{y}) = V_{\mathcal{y}} = (V \otimes V)(I_{m^2} + I_{(m,m)}) = \mathcal{X}P(D \otimes D)P\mathcal{X}',$$

as at Section 3.4, equation [2].

Beginning again, consider the r.v. U = Myy' and an equivalent model:

$$E(U) = E(Myy'M) = MZDZ'M,$$

with covariance operator $\mathcal{V}$ such that $\mathcal{V}(A) = 2V(A)V$ for $A \in \text{Sym}(n)$. To see the equivalence use the vec operator:

$$\text{vec}[E(U)] = \text{vec}(MZDZ'M) = (MZ \otimes MZ)\text{vec}(d) = \mathcal{X}\text{vec}(D),$$

and, since $(I_{m^2} + I_{(m,m)})\text{vec}(A) = 2\text{vec}(A)$ for any $A \in \text{Sym}(n)$,

$$\text{vec}[\mathcal{V}(A)] = \text{vec}(VAV) = (V \otimes V)\text{vec}(A)$$

$$= \tfrac{1}{2}(V \otimes V)(I_{m^2} + I_{(m,m)})\text{vec}(A) = \tfrac{1}{2}V_{\mathcal{y}}\ \text{vec}(A).$$

(See Eaton [1983, Chapter 2 and 4] for a discussion of this view of the covarance as an operator, $\mathcal{V}$. There with the inner product $(A, B) = \text{tr}(AB)$ for A, B in Sym(n), one can write $\text{cov}[(A, U), (B, U)] = 2\text{tr}(AVBV) = 2(A, \mathcal{V}(B))$.)

Since we have seen that estimates quadratic in the data y are linear in the r.v. $\mathcal{y}$, we can apply Theorem 5.5.1 immediately if we can show that $V_{\mathcal{y}}\mathcal{X}_* = \mathcal{X}_*Q$, for some matrix Q. This, in turn, will hold if and only if

$$V_{\mathcal{y}}\mathcal{X}_*\zeta = \mathcal{X}\zeta_1, \text{ for all } \zeta, \text{ and some } \zeta_1 \text{ possibly dependent on } \zeta.$$

This means

$$(V \otimes V)(I_{m^2} + I_{(m,m)})[\text{vec } M_1 \mid \ldots \mid \text{vec } M_k]\zeta$$

$$= [\text{vec } M_1 \mid \ldots \mid \text{vec } M_k]\zeta_1$$

or,

$$(V \otimes V)(I_{m^2} + I_{(m,m)})[\text{vec}(\Sigma\, \zeta_i M_i)] = \text{vec}(\Sigma\, (\zeta_1)_i M_i)$$

where $\Sigma\, \zeta_i M_i$ is now an arbitrary element a, say of $Sp(\sigma^2)$, and $\Sigma\, (\zeta_1)_i M_i = b \in Sp(\sigma^2)$. Thus

$$(V \otimes V)(I_{m^2} + I_{(m,m)})\text{vec}(a) = \text{vec}(b),$$

or

$$2(V \otimes V)\text{vec}(a) = \text{vec}(b),$$

since vec(a) is symmetric. Hence

$$\text{vec}(VaV) = \text{vec}(c), \text{ for } 2c = b,$$

and this is to hold for all $V \in Sp(\sigma^2)$. This is exactly the requirement then that $Sp(\sigma^2)$ be quadratic, so by Theorem 5.5.1 we get Seely [1971] above, and a little more: the OLS estimates are in fact Gauss-Markov, and hence best among a slightly wider class of estimates.

Using the Second Linearization, the condition of quadraticness can be readily seen to be equivalent to the $V_{\mathcal{Y}}$-invariance of Λ_2, that portion of the constraint matrix Λ which is estimable with respect to $\mathcal{X}$, since $V_{\mathcal{Y}}$ has the form $V_{\mathcal{Y}} = \mathcal{X}PQP\mathcal{X}'$, so that $\mathcal{X}$ is always $V_{\mathcal{Y}}$-invariant. The estimable constraints appearing in the Second Linearization are thus shown to be a key to to examining when every estimable function has a Gauss-Markov estimate.

An important example for which the quadratic criterion applies is that of balanced data with arbitrary γ, as recently shown by Anderson et al. [1984]. They also give what is evidently the most rigorous definition to date of exactly what is meant by "balanced" data and this is of significant value in itself. We leave these details to the reader.

Other work along the lines of Seely [1971] is that of Drygas [1977], in that conditions are found for every estimable function to have an optimal estimate, in the sense of Theorem 5.6.1, given that the kurtosis is non-zero. These results are restricted versions of the general solution given in the next chapter: see the comments following Theorem 6.5.1.

Having now discussed the consequences of $Sp(\sigma^2)$ Jordan for quadratic estimation we state the important result of Seely [1971, 1972, 1977] on sufficiency and completeness, and UMVUE's, given normal data.

We begin with a slight modification of our basic model setup. Thus whereas we have worked with estimates $y'by$ with b such that $b = MbM$, we can almost as easily work with estimates $y'cy$ with c such $c = K'cK$ for K such that $M = KK'$, $K'K = I_t$, $t = r(M)$. That such a K exists follows from the spectral decomposition of M; that translation invariance persists can be checked. We have generally chosen not to use this estimate form for reasons of convenience, while now we must use it for a very practical reason: if y has a normal density then so does $K'y$, while in general My does not. This prompts the following:

Definition. For K of full column rank such that $M = KK'$, $K'K = I_t$, $t = r(M)$, let

$$N_i = K'Z_iZ_i'K = K'V_iK, \ 1 \le i \le k,$$

and write

$$Sp_* = \{ \Sigma\, t_i N_i \mid \text{all real } t_i \}.$$

It can be checked that $Sp(\sigma^2)$ is a "quadratic subspace" if and only if Sp_* is so, and thus we can state:

<u>Theorem 5.6.2.</u> For normal data y having covariance $s \in Sp(\sigma^2)$, if $Sp(\sigma^2)$ is a Jordan algebra ("quadratic subspace"), then the r.v.'s $y'N_iy$ are jointly a complete sufficient statistic for σ^2. Moreover, any r.v. ϕ which is a measurable function of the $y'N_iy$ is the uniform minimum variance unbiased invariant estimate (I-UMVUE) of its expectation $E(\phi)$.

<u>(Outline of Proof).</u> Begin by assuming that the N_i are linearly independent; this is possible in the argument to follow as can be checked. Next observe that the density of $z = K'y$ is proportional to $\exp(A)$ for $2A = y'(s^{-1})y$ with $s = \Sigma\, \sigma_i^2 N_i$. With $Sp(\sigma^2)$ Jordan we know $s^{-1} \in Sp_*$: see Section 9.2. Then $s^{-1} = \Sigma\, \theta_i(\sigma^2)N_i$ for functions θ_i, and $\theta = (\theta_1, \theta_2, \ldots, \theta_k)$ is one-to-one, and onto the set of all coefficients $c = (c_1, c_2, \ldots, c_k)$ such that $\Sigma\, c_i N_i$ is invertible. Next, a continuity argument using the determinant shows that the range of θ contains a k-dimensional rectangle. Then the Lehmann-Scheffé result on completeness, and the factorization theorem on sufficiency (see Lehmann [1959], Sections 2.7 and 4.3) together in the usual way show that the r.v.'s $y'\theta_i N_i y$ are jointly sufficient and complete. Finally, Rao-Blackwell (see Bickel and Doksum [1977], Section 4.2) shows that ϕ is the invariant UMVUE of $E(\phi)$. ∎

Chapter Six: The General Solution to Optimal Unbiased Estimation

6.1 Introduction.

In this chapter we bring to a close the first phase of our study, namely the solution to the problem of finding estimates of the components which are unbiased translation invariant, and optimal in the sense of then having minimum variance. It is here that we also introduce structural ideas that are taken up more completely in the following chapters.

As discussed in the introduction to Chapter Five we may view the solution as the appropriate local version of the Seely-Zyskind results, a localization which in fact results in a substantially simpler proof of this most general result.

6.2 A Full Statement of the Problem.

To focus our discussion and make this chapter more self-contained we begin with a complete statement of the original problem.

Thus let $\mathcal{F}$ be the class of distributions having finite first four moments, with covariance structure as described in the basic variance components model of Section 1.3. Let y be a r.v. having distribution $F \varepsilon \mathcal{F}$, and let $p(\mathcal{F})$ be a linear functional on the class $\mathcal{F}$ of the form $p(F) = p'\sigma^2$, that is a linear parametric function of σ^2.

Consider the class $\mathcal{U}$ of r.v.'s which are quadratic in y and also translation invariant. Further, let $\mathcal{U}(p)$ be the subclass $\subseteq \mathcal{U}$ of estimates which are unbiased for a given $p'\sigma^2$. Then an element of $\mathcal{U}(p)$ is of the form $y'ay$ for matrix a with $a = MbM$, b symmetric, and a such that $tr(a, M_i) = p_i$ for all i. Note that we are now making a shift to the more algebraically natural style of using lower case letters for most matrices, a, b, etc., but choose to retain upper case for the matrices which define the model, that is the M_i.

Our problem now is to find, if it exists, that element h(y) ε $\mathcal{U}$(p) having minimum variance in $\mathcal{U}$(p) at every choice of F ε $\mathcal{F}$. That is, we seek h such that at every F ε $\mathcal{F}$, we have have var[h(y)] $\leq$ var[(g(y)] for all g(y) ε $\mathcal{U}$(p). We call such a solution an <u>optimal unbiased estimate</u>.

6.3 The Lehmann-Scheffé Result.

In order to derive our solution we start with a classical result (see Rao [1973a, p. 317]):

<u>Theorem 6.3.1. (Lehmann and Scheffé [1950])</u> For r.v. z, parameter θ, and a linear space $\mathcal{E}$ of estimates based on z, an estimate g(z) ε $\mathcal{E}$, unbiased for θ, is of minimum variance in the subspace of estimates unbiased for θ if and only if

$$\text{cov}[g(z), r(z)] = 0$$

for all r(z) ε $\mathcal{E}$ which are unbiased for zero.

This elegant, broadly applicable result has a specialization which we have already encountered, that is Lemma 5.4.3, as one can readily check, and since it deals with a single estimate at a time it has the key local character we require for our solution to the component estimation problem. Alternatively, it is the force behind the Seely-Zyskind results themselves, as they note (Seely and Zyskind [1971]), and we choose to start from it in order to underline the essential simplicity of our general results.

Note that it is possible to apply the theorem with $\mathcal{E}$ = $\mathcal{U}$(p) since the class $\mathcal{U}$(p) is a real linear space as one can check.

6.4. The Two Types of Closure.

Our beginning is:

Definition. A matrix $b \in Sp(\sigma^2)$ is said to be jordan closed (j-closed) if $aba \in Sp(\sigma^2)$ for all $a \in Sp(\sigma^2)$.

In the definition "jordan" is used since we later show that the space of all such elements is a Jordan algebra (in fact, a special semisimple finite dimensional Jordan algebra).

Recall how we have defined the set of integers $N[j]$ in Section 3.5: it is the collection $\{ (\sum_{i<j} c_i) + 1, \ldots, (\sum_{i<j} c_i) + c_j \}$; we also need to use the structure matrices of Section 3.5: for e_i = the ith unit vector in $\mathbb{R}_m$, write $e_{(i)} = \text{diag } e_i$, and let $s_i = MZe_{(i)}Z'M$, for $1 \leq i \leq m$. Then

Definition. A matrix $b \in Sp(\sigma^2)$ is said to be structurally closed (s-closed) if for all j, $1 \leq j \leq m$:

$$\sum_{i \in N[j]} s_i(b)s_i \quad \in \quad Sp(\sigma^2)$$

It is convenient to have alternative forms of these definitions and we state some of these as lemmas:

Lemma 6.4.2. [1] b is j-closed if and only if $abc + cba \in Sp(\sigma^2)$ for all a and c in $Sp(\sigma^2)$;

[2] b is j-closed if and only if $M_i b M_j + M_j b M_i \in Sp(\sigma^2)$ for all i, j.

Proof. For [1] let $a = c + d$, with c and d any members of $Sp(\sigma^2)$. Then if $aba \in Sp(\sigma^2)$ it follows that $(c + d)b(c + d) \in Sp(\sigma^2)$, so that cbc + cbd + dbc + dbd ε Sp(σ^2). As cbc and dbd are ε Sp(σ^2), we get one direction of [1]. For the other direction simply put a = c in abc + cba.

For [2] recall that Sp(σ^2) is spanned by the M_i, so linearity finishes the proof. ∎

Next recall that Γ is the matrix containing the kurtosis parameters in the expression for var($\mathbf{y}$) in Theorem 3.3.1. Then

Lemma 6.4.3. An element b ε Sp(σ^2) is s-closed if and only if

$$\mathfrak{X}\Gamma\mathfrak{X}'(\text{vec } b) = \text{vec } c,$$

for some c ε Sp(σ^2), and all Γ, where c may be a function of b and Γ.

Proof. The proof follows from Lemma 3.5.1 [4]. Thus we always have

$$\mathfrak{X}\Gamma\mathfrak{X}'(\text{vec } b) = \text{vec}[\sum_{j=1} \sum_{i\varepsilon N[j]} \sigma_j^4 \gamma_j s_i(b) s_i]$$

$$= \text{vec}[\sum_{j=1} r_j \{\sum_{i\varepsilon N[j]} s_i(b) s_i\}]$$

for $r_j = \sigma_j^4\gamma_j$. Now if b is assumed s-closed then the innermost sum on the right is in Sp(σ^2), so the RHS is vec(c) for c ε Sp(σ^2). Conversely, if the RHS is so for all Γ, hence for all r_j, then it is so for $r_t = 1$, $r_j = 0$ if $j \neq t$. But then

$$\text{vec } c = \text{vec } (\sum_{i\varepsilon N[k]} s_k(b) s_k)$$

where c ε Sp(σ^2), and this holds for all t, $1 \le t \le k$. Thus b is s-closed. ∎

6.5. The General Solution.

Everything is in place for our main result.

Theorem 6.5.1. An estimable function $p'\sigma^2$ has a uniformly minimum variance unbiased quadratic translation invariant estimate if and only if the estimate is the OLS $y'hy$ with h both j-closed and s-closed.

Proof. (⇒). Using the Lehmann-Scheffé result Theorem 6.3.1, if $y'hy$ is to be optimal as above then we must have

$$\mathrm{cov}(y'ay, y'hy) = 0$$

for all $a \in Sp(\sigma^2)$ of the form $a = MbM$, $b \in Sp(\sigma^2)$, and such that $(a, M_i) = 0$, that is, for all $a \in Sp(\sigma^2)^{\perp}$.

Using the derivation of Theorem 3.3.2 we must now have

$$[1] \qquad 0 = 2\mathrm{tr}[asbs] + (\mathrm{vec}\ a)'\mathfrak{X}\Gamma\mathfrak{X}'(\mathrm{vec}\ b)$$

for all elements $s \in Sp(\sigma^2)$ and all $a \in Sp(\sigma^2)^{\perp}$.

This must hold now at every $F \in \mathcal{F}$, that is, for all σ^2 and all γ. In particular it must be so for all F with $\gamma = 0$. Then since the product $asbs$ is not a function of γ, we get

$$0 = \mathrm{tr}(ashs),$$

for all $a \in Sp(\sigma^2)^{\perp}$. Hence only $shs \in Sp(\sigma^2)$ is possible, and this for all $s \in Sp(\sigma^2)$. Thus from the definition we see that h must be j-closed.

To complete the direction (⇒), begin with the fact that h is j-closed for all γ, so

$$0 = \mathrm{tr}(ashs)$$

for all $a \in Sp(\sigma^2)$ implies

$0 = (\text{vec } a)'\mathfrak{X}\Gamma\mathfrak{X}'(\text{vec } h)$

as well, for all Γ. Consequently $\mathfrak{X}\Gamma\mathfrak{X}'(\text{vec } h) = \text{vec } d$ for some $d \in Sp(\sigma^2)$, with d possibly a function of Γ (and h). As seen earlier though, from Lemma 6.4.3, this means exacly that h is s-closed.

Note that since h is j-closed h is necessarily $\in Sp(\sigma^2)$ and so is the OLS for the function $p'\sigma^2$.

Finally, as all the steps above are reversible, the direction ($\Leftarrow$) follows. ∎

Because of this result we are now entitled to make the definitions:

Definition. Let $\mathscr{B}^+$ be the space of all elements $b \in Sp(\sigma^2)$ which are both j-closed and s-closed.

Definition. Let $\mathscr{B}$ be the space of all elements $b \in Sp(\sigma^2)$ which are j-closed.

Definition. Given the quadratic estimate y'by we call the matrix b its kernel, and so refer to $\mathscr{B}^+$ as the space of optimal kernels.

Thus $0 \subseteq \mathscr{B}^+ \subseteq \mathscr{B} \subseteq Sp(\sigma^2)$. Further, if it known for data of interest that $\gamma = 0$ then working through the proof of Theorem 6.5.1 shows that $\mathscr{B}^+ = \mathscr{B}$, and every j-closed element provides an optimal unbiased estimate, and conversely.

We also mention that earlier in Kleefe and Pincus [1974], for normal data, our condition of optimality, $sbs \in Sp(\sigma^2)$ for all $s \in Sp(\sigma^2)$, was obtained, though it was evidently not noted then that in fact $b \in Sp(\sigma^2)$, that is that the space of optimal kernels, $\mathscr{B}$, is contained in $Sp(\sigma^2)$, so all optimal unbiased estimates are then just OLS estimates.

Next, we point out a collection of other precursors in the literature: Drygas [1980, 1985], Khatri [1978]. In essence, these authors obtained half of our Theorem 6.5.1, in that equation [1] of the proof was reached but γ was not allowed to vary. This amounts to working with a restricted version of our estimation problem, one which this author feels is an unjustified confounding of local and uniformly best estimation.

Finally, a slight strengthening of the result occurs with the assumption of zero kurtosis. For $T = [tr(M_iM_j)]$ let $\mathcal{P}_T$ be $\mathcal{P}_{\mathcal{C}}$, for $\mathcal{C} = \mathcal{C}(T)$. Then:

<u>Corollary 6.5.2.</u> If the data has kurtosis $\gamma = 0$, then y'by is of minimum variance among all quadratic translation invariant estimates of $p'\sigma^2$ having minimum norm bias if and only if $b \in \mathcal{B}$, and $E(y'by) = [\mathcal{P}_T(p)]'\sigma^2$.

<u>Proof.</u> Essentially, we show that every minimum norm bias estimate of $p'\sigma^2$ is an unbiased estimate of some other estimable function $u'\sigma^2$, so that the theorem can be applied to optimal unbiased estimation of $u'\sigma^2$.

Thus given $p \in \mathbb{R}_k$, consider its decomposition as $p = u + v$, where u is in the range of $T = [tr(M_iM_j)]$, and v is in the Euclidean inner product complement, with $u'v = 0$, $Tv = 0$, $u = \mathcal{P}_T(p)$. Then

$$\begin{aligned}\Sigma\,[p_i - tr(bM_i)]^2 &= \Sigma\,[v_i + u_i - tr(bM_i)]^2 \\ &= \Sigma\,[v_i]^2 + \Sigma\,[u_i - tr(bM_i)]^2 \\ &\quad - 2\,\Sigma\, v_i[u_i - tr(bM_i)].\end{aligned}$$

But

$$\Sigma\, v_iu_i = v'u = 0,$$

and

$$\Sigma\, v_i[\mathrm{tr}(bM_i)] \;=\; \Sigma\, \mathrm{tr}(v_i bM_i) \;=\; \mathrm{tr}[b(\,\Sigma\, v_i M_i)],$$

and if $b = \Sigma\, b_i M_i$, and $Tv = [Tv_{(j)}]$,

$$= \sum_i \sum_j \mathrm{tr}(b_j v_i M_i M_j)$$

$$= \Sigma\, b_j\, \Sigma\, \mathrm{tr}(v_i M_i M_j)$$

$$= \Sigma\, b_j (Tv)_{(j)} \;=\; 0.$$

Hence if $y'by$ is unbiased for $u'\sigma^2$, it will be of minimum bias for $p'\sigma^2$, so choosing $b \in \mathscr{B}$, and such that $u_i = \mathrm{tr}(bM_i)$ solves the problem. ∎

Continuing with the structure theory, we note that the matrix M acts as the identity element on $Sp(\sigma^2)$, and so has a special connection with the space $\mathscr{B}$:

Theorem 6.5.3. [1] $\mathscr{B} = Sp(\sigma^2)$ if and only if $M \in \mathscr{B}$; [2] with $\gamma = 0$, every estimable function has an optimal unbiased estimate if and only if the usual "error" estimate $y'My$ is optimal unbiased for its expectation $\xi'\sigma^2$, with $\xi_i = \mathrm{tr}(M_i)$.

Finally, we use the fact that $\mathscr{B}^+$ is a vector space to obtain an alternative characterization of optimal unbiased estimation. Thus let $\{m_i\}$ be a vector space basis for $\mathscr{B}^+$, $1 \le i \le d = \dim \mathscr{B}^+$, and let b_{ij} be such that

$$m_i \;=\; \Sigma\, b_{ij} M_j, \quad 1 \le i \le d, \quad 1 \le j \le k.$$

Writing $B = [b_{ij}]$ and $\mathscr{X}_+ = \mathscr{X}B' = [\mathrm{vec}\ m_1 \mid \mathrm{vec}\ m_2 \mid \ldots \mid \mathrm{ved}\ m_d]$, we can state:

Theorem 6.5.4. A parametric function, $q'\sigma^2$, of the components has an optimal unbiased estimate if and only if it is "estimable" with respect to the model $E(\mathbf{y}) = \mathcal{X}_+\zeta$, that is, if and only if $(Bq)'\zeta$ is estimable with respect to the model, and this occurs if and only if $Bq = \mathcal{X}_+'\mathcal{X}_+t = BTB't$, for $t \in \mathbb{R}_d$, and $T = [\mathrm{tr}(M_iM_j)]$.

6.6. An Example.

We give an example which generalizes that of LaMotte [1976], showing that in the general one-way random model optimal unbiased estimates do not usually exist. Our result Theorem 6.5.1 extends the result of LaMotte in that arbitrary kurtosis γ is now permitted.

Thus for integer t let 1_t be a t-vector of all 1's let $J_t = (1_t)(1_t)'$ and write 0_t for a $t \times t$ matrix of all 0's. For integers u, v, and w put $n = u + v + w$. Next let

$Z_1 = 1_u \oplus 1_v \oplus 1_w$ (block diagonal)

$Z_2 = I_n$

$M_1 = Z_1Z_1' = J_u \oplus J_v \oplus J_w$

$M_2 = M = I_n$

Consider the model $E(y) = 0$ with

$$\mathrm{var}(y) = \sigma_1^2M_1 + \sigma_2^2M_2 = \sigma_1^2M_1 + \sigma_2^2I_n.$$

Then the structure matrices are

$s_1 = J_u \oplus 0_v \oplus 0_w$

$s_2 = 0_u \oplus J_v \oplus 0_w$,

$s_3 = 0_u \oplus 0_v \oplus J_w$.

Letting as usual e_i = the ith unit vector in $\mathbb{R}_n$, and $e_{(i)} = \mathrm{diag}\ e_i$, we see

$$s_{(i+3)} = e_{(i)}, \text{ for } 1 \leq i \leq n.$$

In our standard model notation we have $k = 2$, $c_1 = 3$, $c_2 = n$, and $m = n + 3$, and with this setup:

[1] $u = v = w \Rightarrow$ every $b \in Sp(\sigma^2)$ is both j- and s-closed. Hence $\mathscr{B}^+ = \mathscr{B} = Sp(\sigma^2)$;

[2] $u = v < w \Rightarrow$ only $h = h_1M_1 + h_2I_n$ with $h_1 = (u + w)h_2$ are s-closed, while no non-zero h is j-closed. Hence $0 = \mathscr{B}^+ = \mathscr{B} \subset Sp(\sigma^2)$;

[3] $u < v < w \Rightarrow$ there are no non-zero j-closed or s-closed elements. Hence $0 = \mathscr{B}^+ = \mathscr{B} \subset Sp(\sigma^2)$.

Finally, we point out that in our notation the work of Anderson et al. [1984], shows that $\mathscr{B}^+ = \mathscr{B} = Sp(\sigma^2)$ for all models having <u>balanced data</u>.

Chapter Seven: Background from Algebra

7.1 Introduction.

In this chapter we pull together the essential parts of rings and algebras that we later use for the structure theory of variance component estimation. We have made an effort to keep the definitions to the minimum needed for an adequate description of this structure theory. Examples have been collected from the statistical literature which have a nontrivial algebraic content, but which may not have been appreciated as such by many readers.

The central theme here will be the conceptual simplicities that follow from showing that one ring or algebra is isomorphic to another, so that seemingly quite diverse objects are made to reveal their real similarities. Thus our structure theory of variance components will be later shown to depend on smaller, well-known objects, so any mixed model involving the components can be classified via an isomorphism and studied in terms of these smaller, better-behaved units.

A core list of references for the material presented is: (general) van der Waerden [1970], Curtis and Reiner [1962, 1981]; (ring theory) Herstein [1968], McCoy [1964]; (σ-rings of sets) Taylor [1966]; (Jordan rings) Jacobson [1968], Herstein [1969].

Concerning proofs, we generally omit them, particularly the longer ones, those usually associated with larger, key results. On the other hand we will include several proofs of the simpler results, in an effort to both give some of the flavor of a topic, and to help fix the definitions in the mind of the reader. Concerning examples, similar rules will apply, with most of them being given as implicit exercises.

In no sense should this chapter be construed as a condensed text on the subjects discussed, but rather as a minimal weaving together of the essential ideas and issues needed to meaningfully discuss the ring theory aspects of variance components.

7.2 Groups, Rings, Fields.

The notion of a group has played a prominent role in inference for some time, specifically in the study of statistical problems that have an inherent symmetry, for which we then seek invariant solutions. Invariance is assumed to be with respect to a given group, and optimal inference becomes a study of group orbits and maximal invariants. Details of this theoretically rich and far-reaching subject can be found in Lehmann [1959] and elsewhere.

The algebraic objects we will study may already be familiar to some readers. However, we do not assume familiarity, and will instead provide complete definitions throughout.

We begin with a formal definition of a <u>group</u>: consider a set $\mathcal{G}$ on which an operation $\circ$ is defined, so that for elements a and b ε $\mathcal{G}$, we have (1) $a \circ b \in \mathcal{G}$, (2) $(a \circ b) \circ c = a \circ (b \circ c)$, (3) an element, 1, called the identity of $\mathcal{G}$, such that $1 \circ a = a \circ 1 = a$ for all $a \in \mathcal{G}$, (4) for every $a \in \mathcal{G}$ there exists an element $a^{-1} \in \mathcal{G}$ such that $a \circ (a^{-1}) = (a^{-1}) \circ a = 1$.

Note that the definition makes no statement about the exact nature of the operation itself, and this generality is needed later when we will have many possible and useful operations available on the same set.

Familiar examples of groups are:

[1] The integers with operation " + "; the integers modulo a prime p;

[2] The set of all real translations of a data vector in $\mathbf{R}_n$.

[3] The set of all n × n matrices having complex valued entries, which are invertible, with operation " × " being the usual matrix product. This is called the general linear group of order n (over $\mathbf{C}$), and is written $\mathcal{GL}(n, \mathbf{C})$.

As an aside we observe that while groups and invariants are highly visible in statistics, it does not appear, at least to this reader, that much use is ever made of the internal structure of a given group. This seems curious, given the depth of the subject of group theory, and given that valuable scientific applications of a notion as elementary as "subgroup" are routinely made in the sciences, particularly in modern physics.

The next object that is still somewhat familiar to many statisticians is that of <u>field</u>. Thus the real and complex numbers are fields, with the formal definition: given a set $\mathcal{F}$ we suppose there is defined for it two operations " + ", and " × ", such that (1) with respect to the operation " + ", $\mathcal{F}$ is a commutative group, so that $a + b = b + a$ for all $a, b \in \mathcal{F}$; (2) with respect to the operation " × ", the non-zero elements of $\mathcal{F}$, $= \mathcal{F}^*$ say, also form a commutative group; (3) the two operations interact as follows: $(a + b) \times c = a \times c + b \times c$.

As an important example of a field consider the set of 2 × 2 matrices of the form:

$$\begin{bmatrix} a & b \\ -b & a \end{bmatrix}$$

with a, and b both real, $\in \mathbf{R}$. This is of interest since one can show that it is structurally identical to the field of complex numbers $\mathbf{C}$. The identification is by means of the mapping which sends the matrix above to $a + bi \in \mathbf{C}$, and this is an example of a <u>field isomorphism</u>:

given two fields $\mathcal{F}_1$, $\mathcal{F}_2$, an isomorphism ϕ is a mapping from $\mathcal{F}_1$ to $\mathcal{F}_2$ such that (1) ϕ is onto, (2) ϕ is 1 - 1, so $\phi(a) = 0$ if and only if $a = 0$, (3) $\phi(a + b) = \phi(a) + \phi(b)$, (4) $\phi(a \times b) = \phi(a) \times \phi(b)$, and (5) $\phi(1) = 1$. With this isomorphism we see that all the field properties of $\mathcal{F}_1$ are present in and derivable from those of $\mathcal{F}_2$, so analysis on $\mathcal{F}_1$ is equivalent to that on $\mathcal{F}_2$. The structure theory for variance components developed later on will make use of just such mappings, so that apparently difficult objects assume a more familiar and tractable form, without any loss of inner structure.

Groups and fields have come together in statistics in the design of experiments. Here the groups are usually finite abelian (commutative) groups and the fields are finite as well (see John [1971, p. 277 et seq.]). The fields here are the Galois fields $GF(p^n)$, having p^n elements (see Lang [1965, p. 182]). For $n = 1$ we have the alternative notation $GF(p) = Z_p$.

The next object we consider is a ring, which lies in a sense midway between a group and a field. Curiously though it has made essentially no appearance in statistics, with one important exception (which more properly belongs to the subject of probability, and which is discussed below). Thus we make the definition: a <u>ring</u> is a set with two operations, " + ", and " × ", such that (1) it is a commutative group under " + ", (2) products $a \times b$ are defined such that distributivity holds, so $(a + b) \times c = a \times c + b \times c$. Note that we <u>do not</u> assume that non-zero elements have inverses, or that products commute (so $a \times b$ would equal $b \times a$). However we will usually add the condition that our rings be unital so that there is an identity element, 1, with $a \times 1 = 1 \times a = a$ for all elements a in the ring. We also make the standard assumption that "ring" unadorned means "associative ring" so that

$$(a \times b) \times c = a \times (b \times c)$$

As a first example, consider a ring of subsets (events) of a class of sets (the sample space). This is a collection of subsets $\mathcal{R}$ contained in $\mathcal{S}$ say, such that (1) the empty set and $\mathcal{S}$ are $\varepsilon\ \mathcal{R}$, (2) $\mathcal{R}$ is closed with respect to " + " defined as $a + b = a \Delta b$ = the symmetric difference $= [a \cap (b^-)] \cup [b \cap (a^-)]$, for a^-, b^- the complements of a and b, and (3) $\mathcal{R}$ is closed with respect to " × " defined as $a \times b = a \cap b$. One can show that $\mathcal{R}$ is an associative ring. Extending this construction by assuming that $\mathcal{R}$ is closed under inverses (= set complements) we get a field of sets. The standard σ-field of probability is now obtained by adding the condition that $\mathcal{R}$ be closed under countable unions, where unions are defined in terms of the operation " + " and " × " as $a \cup b = (a \Delta b) \Delta (a \cap b)$. $\mathcal{R}$ is then an algebra over the trivial finite field of two elements, $= Z_2 = \{0, 1\}$, and so is often also called a σ-algebra.

In our algebraic study of variance components we will use several basic types of rings:

[1] An associative ring, as mentioned above;

[2] A non-associative ring. Here the essential example is constructed as follows. Consider the set of $n \times n$ real symmetric matrices with the usual addition, and take as product the operation " × " such that $a \times b = \frac{1}{2}(ab + ba)$, where the products ab and ba are the usual matrix products. Note that while non-associative, the ring is commutative, $a \times b = b \times a$.

[3] A division ring, where the ring is assumed to have a multiplicative identity element, 1, so that $a \times 1 = 1 \times a = a$ always holds, and such that every non-zero element has an inverse. We do not suppose the product to be commutative, and here the essential example is the quaternions over the real numbers. Thus consider all formal sums of the form $a + bi + cj$

+ dk, for a, b, c and d ε **R**, and multiplication defined by $i^2 = j^2 = k^2 = -1$, $ij = k$, $ji = -k$, $jk = i$, $kj = -i$, $ki = j$, $ik = -j$, and extended linearly for all sums as above. One can check however, that the quaternions with complex coefficients do <u>not</u> form a division ring. This can be seen as follows: letting, more generally, $\mathcal{Q}(\mathcal{F})$ be the quaternions with coefficients in a field $\mathcal{F}$, let $q = a + bi + cj + dk$ be any element, and define its <u>conjugate</u> $q^* = a - bi - cj - dk$, and its <u>norm N(q)</u> = qq^*. One finds $qq^* = a^2 + b^2 + c^2 + d^2$, and $N(q_1q_2) = N(q_1)Nq_2)$, for q_1, $q_2 \ \varepsilon \ \mathcal{Q}(\mathcal{F})$. Then if one calls a field <u>formally real</u> if zero cannot be expressed as a nontrivial sum of squares (or equivalently if -1 cannot be so expressed) we find that $\mathcal{Q}(\mathcal{F})$ will be a division ring if and only if $\mathcal{F}$ is formally real. The field **C** is not so, and in $\mathcal{Q}(\mathbf{C})$ with $a_1 = b_1 = \sqrt{-1}$, $a_2 = -1$, $b_2 = +1$ we get $(a_1 + a_2i)(b_1 + b_2i) = 0$, where we must distinguish $i \ \varepsilon \ \mathcal{Q}(\mathbf{C})$ from $\sqrt{-1} \ \varepsilon$ **C**.

[4] A matrix ring. Here we mean the familiar set of all $n \times n$ matrices, say, over a field, usually the reals, written $\mathbf{R}_{n \times n}$. One can show that there is a natural mapping between $\mathbf{R}_{nm \times nm}$ and the $n \times n$ matrices with elements in $\mathbf{R}_{m\times m}$, which is in fact a ring isomorphism. Here, as with a field isomorphism we assume the mapping takes the identity onto the identity, and takes only zero to zero. Note that we have thus introduced the idea of a matrix ring with elements which are other than real or complex numbers. Hence we can similarly define the matrix ring of $n \times n$ matrices with elements which are in the division ring of real quaternions, and more generally, for a given ring, $\mathcal{R}$, we let $[\mathcal{R}]_n$ (or just $\mathcal{R}_n$) be the ring of $n \times n$ matrices with entries in $\mathcal{R}$.

Beyond these basic types, we next look at some other important examples of rings.

Consider a set with two operations, " + ", and " . " such that it is a commutative group with respect to " + ", is distributive so $(a + b) . c = a . c + b . c$, and for which the product " . " satisfies the two identities:

$$a . b = b . a$$

$$a^{\cdot 2} . (b . a) = (a^{\cdot 2} . b) . a.$$

where $a^{\cdot 2} = a . a$. Such a non-associative ring is called a <u>Jordan ring</u>, and the example above of n × n real symmetric matrices with product $a . b = \frac{1}{2}(ab + ba)$, is a key case. In fact, since the product " . " is defined by means of the standard associative product of all n × n real (not necessarily symmetric) matrices, it is an example of a <u>special Jordan ring</u>, in that it is isomorphic to an additive subgroup of an associative ring which has the same addition " + ", but which uses this derived product. It is because this ring lives within another, sharing its addition but not its product, that we write carefully the derived product as " . ". Generally then, suppose $\mathcal{R}$ is an associative ring with product ab. Let $\mathcal{R}^+$ be the set of all elements of $\mathcal{R}$, equipping it now with the same addition as $\mathcal{R}$, and with the new product $a.b = \frac{1}{2}(ab + ba)$. This is always a Jordan ring, and the real symmetric matrices are then seen to be a Jordan subring of $([\mathbf{R}]_n)^+$.

Next, consider a ring $\mathcal{R}$ together with a field $\mathcal{F}$ such that elements of the two interact as: αb is an element of $\mathcal{R}$, for $\alpha \in \mathcal{F}$, $b \in \mathcal{R}$, and $\alpha(bc) = (\alpha b)c = b(\alpha c)$, for all $\alpha \in \mathcal{F}$, and all b, $c \in \mathcal{R}$. We say the ring $\mathcal{R}$ is an <u>$\mathcal{F}$-algebra</u>. It may be an algebra with respect to other fields, and may be associative or not. A basic example is again the n × n real matrices, for they form an algebra over the reals. Also, the real quaternions are a real algebra, as are the n × n real symmetric matrices. Note that the complex numbers **C** are also a real algebra, as well as complex one, so that a given ring or algebra may have several characterizations as an "algebra".

For a given $\mathcal{F}$-algebra $\mathcal{R}$, using its additive structure and the action of $\mathcal{F}$ on $\mathcal{R}$, we see that $\mathcal{R}$ is a vector space in the usual sense, and we have available both the subspace structure of $\mathcal{R}$ over the field $\mathcal{F}$, and the vector space dimension defined for $\mathcal{R}$. All the algebras we study will be of finite dimension.

An earlier important statistical use of Jordan algebras occurs in variance component estimation when it is desired to find maximum likelihood estimates: assuming the data is normal, and given knowledge of the mean vector, then explicit closed form solutions of the ML equations for σ^2 are available if and only if the $Z_i Z_i'$ span a Jordan algebra. See Szatrowski [1980], Szatrowski and Miller [1980] and Elbassiouni [1983]. A similar result holds if it is desired to find translation invariant ML estimates: closed form solutions exist if and only if $Sp(\sigma^2)$ is Jordan.

A different kind of algebra can be constructed from any finite group $\mathcal{G}$ and any field $\mathcal{F}$ in the following way. Consider all formal finite sums of elements of $\mathcal{G}$ with coefficients in $\mathcal{F}$: $\Sigma\, \alpha_i g_i$, for $\alpha_i \in \mathcal{F}$, $g_i \in \mathcal{G}$. Defining multiplication as

$$(\Sigma\, \alpha_i g_i)(\Sigma\, \beta_i h_i) = \Sigma\, (\alpha_i \beta_j)(g_i h_j)$$

where the right hand sum is over all i, j, then defines a group algebra over $\mathcal{F}$. An important statistical use of the notion of group algebra occurs in the wholly algebraic treatment of zonal polynomials, functions which are used in multivariate density calculations. Farrell [1985] is the central, wonderfully written though advanced reference for this topic.

Still other rings and algebras can be formed from a given one as follows. In any associative ring $\mathcal{R}$ such that $(2)^{-1}$ exists, we can begin with the same elements and define two new products: the Jordan product (as above) with $a \, . \, b = \frac{1}{2}(ab + ba)$, and the Lie product with $[a, b] = ab - ba$. The first operation induces the <u>Jordan structure</u> of the ring $\mathcal{R}$, and the second is said to induce the <u>Lie structure</u> of $\mathcal{R}$ (see Herstein [1969], Chapter 1). In $[\mathbb{R}]_n$ the symmetric matrices are stable under this Jordan product, so that with a' the transpose of a, etc.

$$(a.b)' = [\tfrac{1}{2}(ab + ba)]' = \tfrac{1}{2}[(ab)' + (ba)']$$

$$= \tfrac{1}{2}(ba + ab) = a.b,$$

while the skew-symmetric matrices are stable under this Lie product, and every element of $[\mathbb{R}]_n$ can be written uniquely as the sum of symmetric and a skew-symmetric matrix: $a = \frac{1}{2}(a + a') + \frac{1}{2}(a - a')$. Here then the Jordan and Lie structures form a useful decomposition of the ring, and more generally the two can jointly help characterize the structure of a ring.

7.3 Subrings and Ideals.

We turn our attention now to looking at sub-objects of some of the structures introduced above. Where no confusion will result we write products simply as: ab.

A subgroup is simply a group contained within a given larger group, with the important proviso that it have the same identity element. We will similarly assume the same is true for for the identity element of a subalgebra or subfield.

An important special case of subring is that of an <u>ideal</u>. We begin with the notion of a one-sided left ideal, and here the subring $\mathcal{I}$, say, contained in the ring $\mathcal{R}$ is assumed to have the property that all products $\{ai \mid a \in \mathcal{R}, i \in \mathcal{I}\}$ are contained in $\mathcal{I}$; similarly for a one-sided right ideal. A <u>two-sided ideal</u>, or simply an <u>ideal</u> is taken as one which is both a left and right ideal. For commutative rings these two notions coincide, as for example in any Jordan ring, since a . b = b . a.

Note that in contrast to subalgebra or subfield we do not assume that ideals contain any identity element, let alone the identity of the ring itself.

The ideals of any ring are its basic components of interest and may in certain happy circumstances, provide a complete description of the ring itself. As an example consider a commutative ring $\mathscr{R}$, and call an ideal $\mathscr{I}$, a <u>prime ideal</u> if it is not all of $\mathscr{R}$ and if ab ε $\mathscr{I}$ implies either a ε $\mathscr{I}$ or b ε $\mathscr{I}$. We show the following: if all the ideals of $\mathscr{R}$ are prime then $\mathscr{R}$ is a field. Thus for $x \neq 0$, we get $x^2 \ \varepsilon \ \mathscr{R}x^2$, and since $\mathscr{R}x^2$ is prime, it follows that $x \ \varepsilon \ \mathscr{R}x^2$, so that $x = wx^2$ for some w ε $\mathscr{R}$. But then $x(1 - wx) = 0$, and since (0) is also a prime ideal only $1 - wx = 0$ is possible. Hence x^{-1} exists.

In the case of an algebra, ideals are assumed to be algebra ideals, so that they are first and foremost subalgebras, rather than just subrings, and for some algebras these notions may be distinct: for the integers Z we find $[Z]_2$ is contained in $[\mathbb{R}]_2$ only as a subring. Thus we assume a subalgebra is a subring with the added property that it is stable under the action of the field, so for subalgebra $\mathscr{B}$ contained in $\mathscr{F}$-algebra $\mathscr{A}$, we have $\alpha\mathscr{B} \subseteq \mathscr{B}$ for all $\alpha \ \varepsilon \ \mathscr{F}$.

As examples of subrings and ideals consider:

[1] the set $\mathscr{I}$ contained in real 2 × 2 matrices, such that every element of $\mathscr{I}$ is of the form:

$$\begin{bmatrix} a & b \\ 0 & 0 \end{bmatrix}$$

This is a subring of the 2 × 2 matrices, but is not an ideal. Also it has an infinite number of left identity elements but no right identity, so that there exists $1_\ell \ \varepsilon \ \mathscr{I}$ for which $1_\ell a = a$ but not necessarily $a1_\ell = a$, for all a ε $\mathscr{I}$. $\mathscr{I}$ is however a right ideal in the ring of 2 × 2 matrices, in fact a right algebra ideal.

[2] Letting $\mathcal{Q} = \mathcal{Q}(\mathbf{R})$ be the division algebra of real quaternions consider the mapping given by

$$a + bi + cj + dk \quad \to \quad \begin{bmatrix} a + bi & c + di \\ -c + di & a - bi \end{bmatrix}$$

Then one can check that this is a real algebra isomorphism of $\mathcal{Q}$ onto a subring of $[\mathbf{C}]_2$, the real algebra of 2 × 2 matrices with complex-valued entries. From this it now follows without further multiplication that $\mathcal{Q}$ is associative, since the usual matrix product is so.

Alternatively consider the mapping

$$a + bi + cj + dk \quad \to \quad \begin{bmatrix} a & -b & -c & -d \\ b & a & d & -c \\ c & -d & a & b \\ d & c & -b & a \end{bmatrix}$$

It can be shown that this defines an isomorphism of $\mathcal{Q}$ into $[\mathbf{R}]_4$.

For an <u>associative</u> ring $\mathcal{R}$ and a set $\mathcal{S}$ of elements of $\mathcal{R}$ consider the set of all sums of the form $\Sigma\, s_i r_i$, over every finite set s_i of elements of $\mathcal{S}$, and r_i of $\mathcal{R}$. This is said to form the <u>right ideal</u> generated by $\mathcal{S}$ in $\mathcal{R}$; if $\mathcal{R}$ has an identity element then the ideal contains $\mathcal{S}$ itself. Similarly for the left ideal generated by $\mathcal{S}$. A more important case for us will be the <u>(two-sided) ideal</u> generated by $\mathcal{S}$: it consists of all sums of the form

$$\Sigma\, s_i r_i \; + \; \Sigma\, r_j s_j \; + \; \Sigma\, r_k s_{k\ell} r_\ell,$$

for all finite sequences s_i, s_j, $s_{k\ell}$ ε $\mathcal{S}$, and r_i, r_j, r_k, r_ℓ ε $\mathcal{R}$. A much more basic new set is formed from $\mathcal{S}$ by taking all finite sums of terms of the form $s_i s_j \ldots s_k$, for s_i, s_j, ..., s_k ε $\mathcal{S}$, and this is called the subring generated by $\mathcal{S}$ in $\mathcal{R}$. If $\mathcal{S}$ should also happen to be a member of a ring $\mathcal{R}_0$ with a product different from that of $\mathcal{R}$, then the corresponding generated subrings will also usually be distinct. This situation will in fact be important to us later when we consider the associative as well as the Jordan product ring generated by a set of matrices in $[\mathbf{R}]_n$, and the precise relation between the two subrings will be an object of study in itself.

A somewhat more abstract but actually more useful definition of a subring generated by set $\mathcal{S}$ is that it is the smallest ring in $\mathcal{R}$ which contains $\mathcal{S}$; alternatively it is the intersection of all subrings of $\mathcal{R}$ containing $\mathcal{S}$. Similarly for the one- and two-sided ideals generated by $\mathcal{S}$.

If for any ring there are no nonzero, proper ideals, and the multiplication is nontrivial, so that $\mathcal{R}^2 \neq 0$, then we say the ring is simple. Similarly for any algebra, where we require all ideals in consideration to be algebra ideals.

Finally, as further examples of the notion of ideal consider:

[1] Letting 2Z and 3Z be the ideals in the integers Z generated by 2 and 3, respectively, we find the (set-theoretic) union 2Z ∪ 3Z is not an ideal, while 2Z ∩ 3Z is. On the other hand 2Z ∪ 4Z = 2Z is an ideal.

[2] If $\mathcal{I}$ is an ideal in ring $\mathcal{R}$, then $[\mathcal{I}]_n$ is an ideal in $[\mathcal{R}]_n$.

[3] Using the family of all subsets $\mathcal{A}$ of a given set $\mathcal{S}$, and the ring operations defined earlier for sets, we can see that if $\mathcal{B}$ is a fixed subset of $\mathcal{S}$, then the set of all subsets $\mathcal{A}$ of $\mathcal{S}$ which contain no element of $\mathcal{B}$, forms an ideal in $\mathcal{S}$.

[4] In any field $\mathcal{F}$ the only ideals are 0 and the field itself, so any field is a simple ring.

Next, two other important methods of forming new rings from old are the sum and product. In each case we begin by assuming that the finite set of rings $\mathcal{R}_i$ are all contained in some larger ring $\mathcal{R}$.

For two rings $\mathcal{R}_1$ and $\mathcal{R}_2$ their sum is defined to be the set of all sums $r_1 + r_2$, over all $r_1 \in \mathcal{R}_1$, $r_2 \in \mathcal{R}_2$, where we assume $\mathcal{R}_1$, $\mathcal{R}_2$ are contained in some larger ring $\mathcal{R}$, so the summation is well-defined.

Next, we say the ring $\mathcal{A}$ is the direct sum of the $\mathcal{R}_i$, $\mathcal{A} = \oplus \mathcal{R}_i$, if and only if every element $a \in \mathcal{A}$ is uniquely expressible as a sum of elements of the $\mathcal{R}_i$, so that $a = \Sigma r_i$, $r_i \in \mathcal{R}_i$. Equivalently, the ring formed by taking all finite sums of elements of the $\mathcal{R}_i$ is a direct sum if the sum is zero only when all its terms are zero.

The product of two subrings $\mathcal{R}$, $\mathcal{S}$, is simply the set of all finite sums of the form $\Sigma r_i s_i$, for $r_i \in \mathcal{R}$, $s_i \in \mathcal{S}$. Order is important, and the product is generally not commutative. For any ring $\mathcal{R}$ we can however define $\mathcal{R}^n$ unambiguously by induction as $\mathcal{R}^n = \mathcal{R}(\mathcal{R}^{n-1})$.

Some essential examples of sums and products of subrings are:

[1] Letting k be the least common multiple of two integers m, n, and d their greatest common divisor, we find $mZ \cap nZ = kZ$, $(mZ)(nZ) = (mn)Z$. Letting (m, n) be the ideal generated by m and n in Z leads to $(m, n) = mZ + nZ = dZ$. Note then that $7Z + 13Z = Z$.

[2] For two (two-sided) ideals $\mathcal{A}$, $\mathcal{B} \subseteq \mathcal{R}$, we always have $\mathcal{A}\mathcal{B} \subseteq \mathcal{A} \cap \mathcal{B}$; if $\mathcal{R}$ is associative and $\mathcal{B}$ a right ideal, then $\mathcal{R}\mathcal{B}$ is a two-sided ideal and for any element $a \in \mathcal{R}$, $\mathcal{R}a$ is a left ideal, $\mathcal{R}a\mathcal{R}$ a two-sided ideal.

Finally, one more way of getting new rings is needed. This is the ring parallel of Z_p, the integers modulo a prime p. Here one starts with any ring $\mathcal{R}$, associative or not, and an ideal $\mathcal{A} \subseteq \mathcal{R}$. Then we form $\mathcal{R}_{\mathcal{A}} = \mathcal{R}/\mathcal{A}$ as the ring $\mathcal{R}$ where we now take addition and multiplication modulo $\mathcal{A}$. For example, $Z_p = Z/\{pZ\}$. $\mathcal{R}_{\mathcal{A}}$ is called a factor ring of $\mathcal{R}$, or a residue class ring: see McCoy [1964]. Of particular importance is that there is a one-to-one, onto correspondence between the ideals of $\mathcal{R}/\mathcal{A}$ and those ideals $\mathcal{B}$ of $\mathcal{R}$ such that $\mathcal{R} \supseteq \mathcal{B} \supseteq \mathcal{A}$.

7.4 Products in Jordan Rings.

In this section we will follow the presentation of Jacobson [1968, Chapter V].

We adopt the convention that unorganized products

$$a_1.a_2.a_3. \ \dots \ .a_n$$

are taken to mean

$$(\dots((a_1.a_2).a_3). \ \dots \ .a_n),$$

so that the product is formed from left to right:

$$a_1.a_2.a_3 = ((a_1.a_2).a_3).$$

Note that by commutativity a.b = b.a, so that for example,

$$a.b.c = b.a.c = c.(a.b) = c.(b.a).$$

As in any non-associative ring, products of ideals in a Jordan ring or algebra $\mathcal{J}$, must also be handled more carefully, and to study them one begins with the notion of associator ideal. Thus we say the subspace $\mathcal{B} \subseteq \mathcal{J}$ is an associator ideal if the triple product

$$[a, b, c] = (a.b).c - a.(b.c) = a.b.c - c.b.a.$$

is in $\mathcal{B}$ if any one of a, b, or c is in $\mathcal{B}$.

We see that any ideal in $\mathcal{J}$ is already an associator ideal, and if a, b, c ε $\mathcal{J}$ then

$$[a, b, c] = -[c, b, a]$$

and one can check that

$$[a, b, c] + [b, c, a] + [c, a, b] = 0.$$

Hence a subspace $\mathcal{B}$ is an associator ideal if and only if $[b, a_1, a_2] \varepsilon \mathcal{B}$ for for all b ε $\mathcal{B}$, and a_1, a_2 ε $\mathcal{J}$. Next, given $\mathcal{B}$ an associator ideal, the subspace $\mathcal{B} + \mathcal{B}.\mathcal{J}$ is an ideal, since b ε $\mathcal{B}$ and a_1, a_2 ε $\mathcal{J}$ implies $b.a_1.a_2 = a_1.a_2.b + [b, a_1, a_2] \varepsilon \mathcal{B}.\mathcal{J} + \mathcal{B}$.

If the subrings $\mathcal{B}_1$ and $\mathcal{B}_2$ are ideals (associator ideals) then $\mathcal{B}_1 \cap \mathcal{B}_2$ and $\mathcal{B}_1 + \mathcal{B}_2$ are ideals (associator ideals), while their product $\mathcal{B}_1.\mathcal{B}_2$ need not be. Nonetheless we can quote from Jacobson [1968, Lemma 1, p. 190]:

Lemma 7.4.1. If $\mathcal{B}_1$ and $\mathcal{B}_2$ are ideals then $\mathcal{B}_1.\mathcal{B}_2$ is an associator ideal, and $\mathcal{B}_1.\mathcal{B}_2 + \mathcal{B}_1.\mathcal{B}_2.\mathcal{J}$ is an ideal; if $\mathcal{B}_1$, $\mathcal{B}_2$, and $\mathcal{B}_3$ are ideals then $\mathcal{B}_1.\mathcal{B}_2.\mathcal{B}_3 + \mathcal{B}_2.\mathcal{B}_3.\mathcal{B}_1 + \mathcal{B}_3.\mathcal{B}_1.\mathcal{B}_2$ is also; if $\mathcal{B}$ is an ideal then $\mathcal{B}^{\cdot 3} = \mathcal{B}.\mathcal{B}^{\cdot 2}$ is also, for $\mathcal{B}^{\cdot 2} = \mathcal{B}.\mathcal{B}$.

7.5. Idempotent and Nilpotent Elements.

With the product of two ideals defined, and in particular, the powers of a single ideal, we look at the powers of single elements, and the relationships between the two. Thus we call a ε $\mathcal{R}$ nilpotent if $a^n = 0$ for some integer n ε Z, and call an ideal nil if every member is nilpotent, $a^{n(a)} = 0$ for every a ε $\mathcal{R}$, with n(a) ε Z in general a function of the element a.

A kind of opposite to nilpotent is idempotent: an element a in a ring $\mathcal{R}$ is idempotent if $a^2 = a$. The identity element of $\mathcal{R}$ is always an idempotent, and many rings have many distinct idempotents. Thus if m and n are relatively prime integers, then in the ring of integers modulo mn, there are at least two idempotents other than 0 and 1; for p a prime, the only idempotents in Z_p are 0 and 1.

As an example using both idempotents and nilpotents consider any associative ring which has no nonzero nilpotent elements. Then we show that every idempotent element commutes with every other element of the ring. Thus consider e idempotent, and x, any other element of the ring. Then

$$[(ex - xe)e]^2 = (ex - xe)e(ex - xe)e$$

$$= (exe - xe)exe - xe)$$

$$= exexe - xexe - exexe + xexe = 0.$$

Since there are no nonzero nilpotent elements, only

$$(ex - xe)e = 0$$

is possible, implying exe = xe. Similarly

$$[e(ex - xe)]^2 = 0$$

leading to ex = exe, hence ex = exe = xe, and commutativity is thus verified.

7.6. The Radical of an Associative or Jordan Algebra.

Having looked at powers of elements we now study powers of ideals in associative or Jordan algebras. This will end with our identifying that part of these algebras that resists easy classification and understanding, while the remaining, tractable part of these algebras will then form the topic of the next chapter.

We begin with an associative ring $\mathcal{R}$ by saying that an ideal $\mathcal{A} \subseteq \mathcal{R}$ is nilpotent if $\mathcal{A}^n = 0$ for some integer n. Certainly, if $\mathcal{A}$ is nilpotent then every element of $\mathcal{A}$ is nil, and it is an important fact that for finite dimensional associative algebras the converse holds: this is Levitzki's Theorem (see Herstein [1968, p. 37]).

For a Jordan algebra, as we might expect, matters are a little more delicate. Thus suppose we define the powers $\mathcal{J}^{\cdot r}$, $r = 2^k$, by:

for $r = 2^0$, $\mathcal{J}^{\cdot r} = \mathcal{J}$; for $r = 2^k$, $\mathcal{J}^{\cdot r} = (\mathcal{J}^{\cdot t})^{\cdot 2}$, for $t = 2^{k-1}$. Then $\mathcal{J}^{\cdot 2}$ is an ideal, by Lemma 7.4.1, and all the powers $\mathcal{J}^{\cdot r}$, $r = 2^k$, are subalgebras but <u>need not be ideals</u>. Still, we say that $\mathcal{J}$ is <u>solvable</u> if $\mathcal{J}^{\cdot r} = 0$ for $r = 2^n$ and some integer n. The basic result on solvable ideals is:

<u>Lemma 7.6.1.</u> If $\mathcal{B}_1$ and $\mathcal{B}_2$ are solvable ideals in $\mathcal{J}$ then so is $\mathcal{B}_1 + \mathcal{B}_2$.

For $\mathcal{J}$ taken to be finite dimensional it then follows immediately that there must exist a solvable ideal $\mathcal{B}$ = <u>rad($\mathcal{J}$) = the radical of $\mathcal{J}$</u> such that $\mathcal{B}$ contains every solvable ideal of J; in this sense rad(J) is the maximal solvable ideal of J. The proof that rad(J) actually exists runs this way. Begin with any nilpotent ideal; if none should exist then rad($\mathcal{J}$) = 0. Otherwise find another nilpotent ideal and form the sum of the two. Continuing this way leads to an increasing chain of ideals, say, $\mathcal{B}_1 \subseteq \mathcal{B}_2 \subseteq \mathcal{B}_3 \subseteq \ldots$. Then because of finite dimensionality this chain must become statioinary at some point, with, say, $\mathcal{B}_{k-1} \subseteq \mathcal{B}_k = \mathcal{B}_{k+1} \ldots$. Here $\mathcal{B}_k$ is exactly rad($\mathcal{J}$). An alternative, equivalent argument is found by using all chains of nilpotent ideals (see Kelley [1955, p. 32]). Finite dimensionality provides a member of the chain containing all the other members, so a maximal nilpotent ideal must exist.

If we agree to call an ideal <u>maximal, with respect to property $\mathcal{P}$ (or just maximal $\mathcal{P}$)</u>, if it is contained in no other proper ideal also having property $\mathcal{P}$, then we can say that rad($\mathcal{J}$) is the maximal solvable ideal of $\mathcal{J}$.

If in any non-associative algebra we agree to call an ideal <u>nilpotent</u> if there exists an integer n such that every product of n elements (associated in any manner) in the ideal is zero, then it follows that if $\mathcal{J}$ is a nilpotent Jordan algebra it is necessarily solvable. The converse is due to Albert, and appears as Corollary 1, p. 195 of Jacobson [1968]:

Theorem 7.6.2. Any finite dimensional solvable Jordan algebra is nilpotent.

Turning now to associative algebras $\mathcal{R}$, we can quote from Curtis and Reiner [1962, p. 161], to the effect that:

Theorem 7.6.3. In any finite dimensional associative algebra $\mathcal{R}$, the sum of any finite number of nilpotent ideals is nilpotent.

Thus in such algebras $\mathcal{R}$ there is similarly defined rad($\mathcal{R}$), commonly called the Jacobson radical of $\mathcal{R}$. It is the sum of all nilpotent ideals in $\mathcal{R}$, and can be shown to be itself a nilpotent ideal, which also contains every nilpotent ideal of $\mathcal{R}$. This last fact shows that in $\mathcal{R}$ associative, rad($\mathcal{R}$) is the maximal nilpotent ideal of $\mathcal{R}$, and prompts the following consolidation: in associative $\mathcal{R}$ or Jordan $\mathcal{J}$ the radical is the unique maximal nilpotent ideal.

If it should happen that rad($\mathcal{R}$) or rad($\mathcal{J}$) = 0 for associative $\mathcal{R}$ or Jordan $\mathcal{J}$, then we say these algebras are semisimple. And yes, there is a strong connection between simple and semisimple: see the next chapter. Also, as a footnote on "semisimplicity" we observe that in the statistical literature this term is reserved for a symmetric matrix whose minimum polynomial factors into distinct linear terms (see Rao and Mitra [1971, p. 10]). Briefly, this definition is exactly equivalent to: the algebra (polynomial ring) generated over $\mathbf{C}$ by the matrix is semisimple, that is, has radical zero. Details are found in Curtis and Reiner [1962, Exercise 7, p. 163.]

We note that for many questions one may reduce to the semisimple case by means of the important fact: the radical of the factor ring $\mathcal{R}/\text{rad}(\mathcal{R})$ is exactly zero (see Herstein [1968, p. 15]).

Let's look at some examples of radicals of rings:

[1] For any commutative, associative ring its radical is the set of all nil elements. Thus for an integer m and Z_m, the integers (modulo m), one finds that $\text{rad}(Z_m)$ is the set of all multiples of the prime factors appearing in the factorization of m. Following up on this, note that while every Jordan algebra is commutative, the argument just used fails to go through because of the lack of associativity; for example, $a.b.a.b \neq (a^{\cdot 2}).(b^{\cdot 2})$, in general.

[2] Consider $[\mathbb{R}]_2$, the algebra of real matrices. One can check that $[\mathbb{R}]_2$ has no nonzero, proper ideals, so that $\text{rad}([\mathbb{R}]_2) = [\mathbb{R}]_2$ or $= 0$. But the algebra itself is not nilpotent (it has an identity element!) so only $\text{rad}([\mathbb{R}]_2) = 0$ is possible. Note though that the algebra contains nilpotent one-sided ideals, for example the set of all matrices which are zero but for the upper right corner.

[3] This next result is <u>Maschke's Theorem</u> (see Herstein [1968, p. 26]) and can be used as an alternative proof of a key result in the study of zonal polynomials (see Farrell [1985, Corollary 12.3.2]; there the critical fact used is that the radical of the real symmetric group algebra is zero): Let $\mathcal{G}$ be finite group and consider the group algebra $\mathcal{A}_{\mathcal{F}}(\mathcal{G})$ of $\mathcal{G}$ over a field $\mathcal{F}$. For any field consider the sum $1 + 1 + \ldots + 1 = n1$ of the identity element n times, and let p be the smallest positive integer such that $p1 = 0$, if there is such a finite integer, and put $p = 0$ otherwise. This p is called the <u>characteristic</u> of the field, $p = \text{char } F$, and is always a prime number (or 0), as can be shown. The theorem is now: for a finite group $\mathcal{G}$ and field $\mathcal{F}$ with $p = \text{char } \mathcal{F}$, if $p = 0$ or if p does not divide the

order of the group (= its number of elements), we have rad $\mathcal{A}_{\mathcal{F}}(\mathcal{G}) = 0$.

We consider now a more extensive example of Jordan rings and radicals, and a special case of it (the radical = 0 case) will appear later as an important ingredient in the complete classification of the semisimple Jordan algebras. Additional details can be found in Jacobson [1968, p. 13-14.]

Thus begin with a vector space $\mathcal{B}$ over $\mathbf{R}$ and assume $\mathcal{B}$ is equipped with a symmetric bilinear form f: for x, y ε $\mathcal{B}$, $f(x, y) \varepsilon \mathbf{R}$, $f(x, y) = f(y, x)$, $f(x + z, y) = f(x, y) + f(z, y)$, and $f(\alpha x, y) = \alpha f(x, y)$, for any $\alpha \varepsilon \mathbf{R}$. Consider $\mathcal{J} = \mathbf{R} \oplus \mathcal{B}$ which is the vector space direct sum of $\mathbf{R}$ and $\mathcal{B}$. We can define a product in $\mathcal{J}$ by

$$(\alpha + x) . (\beta + y) = [(\alpha\beta + f(x, y)) + (\beta x + \alpha y)],$$

with $\alpha, \beta \varepsilon \mathbf{R}$ so $(\alpha\beta + f(x, y) \varepsilon \mathbf{R}$, and $(\beta x + \alpha y) \varepsilon \mathcal{B}$. The symmetry of f implies that the product is commutative, and one can directly check that the other defining relation, $a^{\cdot 2}.(a.b) = a.(a^{\cdot 2}.b)$, of a Jordan algebra in fact holds. We call this $\mathcal{J}$ the <u>Jordan algebra of the symmetric bilinear form f</u>. Next letting $\mathcal{B}^{\perp}$ be the set of of all elements z such that $f(x, z) = 0$ for all x ε $\mathcal{B}$, we find that $\mathcal{B}^{\perp}$ is not only an ideal in $\mathcal{J}$, but that also $\mathcal{B}^{\perp} = \text{rad}(\mathcal{J})$. Thus if f is <u>nondegenerate</u>, in the sense that $f(x, z) = 0$ for some z and all x implies $z = 0$, then $\text{rad}(\mathcal{J}) = 0$ and $\mathcal{J}$ would be semisimple. In fact, if dim $\mathcal{B} > 1$ and f is nondegenerate then $\mathcal{J}$ has no nonzero, proper ideals at all: $\mathcal{J}$ is simple. Finally, one can show that $\mathcal{J}$ is a division algebra (= a division ring and an algebra) if and only if $f(x, x)$ is not a square in $\mathbf{R}$, for any x ε $\mathcal{B}$.

7.7. Quadratic Ideals in Jordan algebras.

We now introduce a type of ideal in a Jordan algebra that is basic to the structure theory of Jordan algebras. These Jordan details can, as usual, be found Jacobson [1968].

To start, we define the Jordan triple product

$$\{abc\} = a\,.\,b\,.\,c\ +\ b\,.\,c\,.\,a\ -\ a\,.\,c\,.\,b$$

for any a, b, and c ε $\mathcal{J}$. If $\mathcal{J}$ is a special Jordan algebra, so that its product a.b is derived from that of an associative algebra containing it, with $a.b = \frac{1}{2}(ab + ba)$, then it is immediate that

$$\{abc\} = \tfrac{1}{2}(abc + cba),$$

where the products abc and cba are the associative products. In particular then, {aba} = aba.

Next define a linear mapping by $x \to \{axb\}$ and denote this by $U_{a,b}$; for a = b we write just $U_{a,a} = U_a$. Note that U_a is quadratic in a, so that if $\mathcal{F}$ is the base field for $\mathcal{J}$ then $U_{\alpha a} = \alpha^2 U_a$, for $\alpha \ \varepsilon\ \mathcal{F}$, a ε $\mathcal{J}$, and also one checks that $U_{a,b}$ is bilinear in a and b, with

$$\tfrac{1}{2}(U_{a+b} - U_a - U_b) = U_{a,b}.$$

A further useful result about U_a, and one much harder to prove in general, is:

$$U_a U_b U_a = U_c, \quad \text{for} \quad c = bU_a,$$

for all a and b ε $\mathcal{J}$. Note that for $\mathcal{J}$ special, $\mathcal{J} \subseteq \mathcal{R}^+$, for $\mathcal{R}$ associative, the result follows from associativity of the product in $\mathcal{R}$.

We can now define a quadratic ideal of $\mathcal{J}$ to be a subspace $\mathcal{B} \subseteq \mathcal{J}$ such that $\mathcal{J}U_b$ is contained in $\mathcal{B}$ for every b ε $\mathcal{B}$. Using the facts just shown about $U_{a,b}$, we can see that $\mathcal{B}$ is a quadratic ideal if and only if $\{b_1 a b_2\} = aU_{b_1,b_2}\ \varepsilon\ \mathcal{B}$ for all a ε $\mathcal{J}$. If $\mathcal{J}$ is a special Jordan algebra then the condition reduces to bab ε $\mathcal{B}$ for all a ε $\mathcal{J}$ and all b ε $\mathcal{B}$, where the product bab is the associative one of the larger algebra in which $\mathcal{J}$ lives.

Any ideal of $\mathcal{J}$ is a quadratic ideal, and for b_1, b_2, $b_3 \in \mathcal{B}$, a quadratic ideal, we also always have $\frac{1}{2}\{b_1b_2b_3\} + \frac{1}{2}\{b_2b_1b_3\} = b_1.b_2.b_3 \in \mathcal{B}$. Note also that letting $\mathcal{R}$ be any associative algebra we see that any one sided ideal of $\mathcal{R}$ is a quadratic ideal of the Jordan subalgebra $\mathcal{R}^+$.

As a further useful observation about quadratic ideals, if $\mathcal{B}$ is a quadratic ideal then we can show that $\mathcal{B}U_a$ is also for any $a \in \mathcal{J}$. Thus if $c \in \mathcal{B}U_a$, then $c = bU_a$ for some $b \in \mathcal{B}$, and $\mathcal{J}U_c = \mathcal{J}U_d$ for $d = bU_a$. But using the important fact given above about U_a, we get $\mathcal{J}U_c = \mathcal{J}U_aU_bU_a$, and this is contained in $\mathcal{B}U_a$. In particular, $\mathcal{J}U_a$ is always a quadratic ideal for any $a \in \mathcal{J}$, and this we choose to call the <u>principal quadratic ideal generated by a</u>.

Other handy properties of quadratic ideals are: If $\mathcal{B}$ is quadratic and $\mathcal{C}$ is any ideal of $\mathcal{J}$ then $\mathcal{B} + \mathcal{C}$ is a quadratic ideal; if we agree to call, as usual, a quadratic ideal <u>maximal</u> if it is properly contained in no proper quadratic ideal of $\mathcal{J}$, then if $\mathcal{D}$ is the intersection of all maximal quadratic ideals of $\mathcal{J}$ we get that $z \in \mathcal{D}$ implies $(1 - z)^{-1}$ exists. Still further, one can show that in any finite dimensional Jordan algebra (with identity element) over an infinite field, rad($\mathcal{J}$) is exactly the intersection of all the maximal quadratic ideals of J (see Jacobson [1968, Exercise 3, p. 202]).

We complete our discussion of quadratic ideals by noting their connection with idempotents, beginning by calling an idempotent $e \in \mathcal{J}$ a <u>primitive idempotent</u> if e cannot be written as the sum $e = e_1 + e_2$ of two nonzero orthogonal idempotents, with $e_1^{\cdot 2} = e_1$, $e_2^{\cdot 2} = e_2$, and $e_1.e_2 = 0$. Next, we call the idempotent <u>absolutely primitive</u> if every element of $\mathcal{J}U_e$, = the principal quadratic ideal generated by e, is of the form $\alpha e + z$, for $\alpha \in$ the field $\mathcal{F}$ of coefficients for the $\mathcal{F}$-algebra $\mathcal{J}$, and z a nilpotent element.

For e an absolutely primitive idempotent, one can now show that the only idempotents in $\mathcal{J}U_e$ are 0 and e, and a basic result in the development of the structure theory of semisimple Jordan algebras which uses these ideas is Theorem 5 of Jacobson [1968, Chapter V]:

Theorem 7.7.1. Let $\mathcal{J}$ be a Jordan algebra such that every element of $\mathcal{J}$ is of the form $\alpha 1 + z$, for $a \in \mathcal{F}$, z nilpotent. Then $\mathcal{J} = \mathcal{F} \oplus \mathcal{N}$, where $\mathcal{N}$ is the set of all nilpotent elements of $\mathcal{J}$ and is itself an ideal of $\mathcal{J}$.

This completes our discussion of the basic algebra background. We move next to the detailed structure of semisimple associative and Jordan algebras.

Chapter Eight: The Structure of Semisimple Associative and Jordan Algebras

8.1 Introduction.

Following the general algebra background work of the last chapter, we are now in a position to specialize our algebras to the kind we will encounter in the structure theory of variance components. Thus we present the complete classification of finite dimensional semisimple associative algebras and also that of finite dimensional semisimple Jordan algebras which are generated by sets of real symmetric matrices. The classification of all the simple Jordan algebras of this form is evidently a new result.

Both of these classifications are very classical results and are very strong results in that the final details are so fully specified and the cases so few in number. Thus for the associative algebras the basic building units are just the complete matrix blocks over $\mathbb{R}$ or $\mathbb{C}$, while in the Jordan setting they are, with one exception, just the set of all "symmetric" matrices over $\mathbb{R}$, $\mathbb{C}$ or the real quaternions.

An important proviso in our classification program is that the isomorphisms postulated by the theorems will not generally be easy to identify in a given practical problem, and in fact this is a research area in computational algebra. Still, a great deal of information can be gained from just the nature of the classification itself, as we shall see in later chapters.

8.2 The First Structure Theorem.

The first result we present is one that applies to finite

dimensional semisimple associative and Jordan algebras. The associative form of the theorem is due to Wedderburn and Artin: see Curtis and Reiner [1968, Theorem 25.15]. The Jordan form is due to Albert: see Jacobson [1968, Chapter V, Section 5].

Theorem 8.2.1. Any finite dimensional semisimple associative or Jordan algebra has an identity element and is a direct sum of algebra ideals which are themselves simple algebras.

We remind the reader that a direct sum of ideals here means that if $\mathcal{A}$ is the algebra, associative or Jordan, then

$$\mathcal{A} = \mathcal{A}_1 \oplus \mathcal{A}_2 \oplus \ . \ . \ . \ \oplus \mathcal{A}_t$$

where the $\mathcal{A}_i$ are algebra ideals, all either associative or Jordan, and every element $a \varepsilon \mathcal{A}$ has a unique expression as a sum of elements from the $\mathcal{A}_i$: $a = \Sigma a_i$, $a_i \varepsilon \mathcal{A}_i$.

Note that while we have consistently assumed our algebras had an identity element, we see that $\text{rad}(\mathcal{J}) = 0$ (semisimplicity) implies in itself the existence of an identity. Also, we can see the meaning behind defining $\mathcal{J}$ to be semisimple if its radical is zero: we get from the above theorem that $\mathcal{J}$ is then "almost" simple, being a direct sum of simple algebras. We'll refer to Theorem 8.2.1 as the First Structure Theorem for semisimple Jordan (or associative) algebras.

Given Theorem 8.2.1 the remaining task is the complete description of all simple finite dimensional associative or Jordan algebras. This we now do, beginning with the Jordan case.

8.3 Simple Jordan Algebras.

The problem of describing all simple Jordan algebras is, from our perspective, a formidable one, and we spare ourselves this effort by making a reduction to the Jordan algebra case that applies in the variance component setting. And to do this, we require the notions of formally real and involution.

Thus we say a Jordan algebra is <u>formally real</u> if

$$a^{\cdot 2} + b^{\cdot 2} = 0 \Rightarrow a = b = 0.$$

for all a and b $\varepsilon\ \mathcal{J}$. Note how this parallels the definition of a formally real field that we introduced in the quaternion discussion of Chapter 7, Section 7.2.

As an example consider Sym(n) = the space of real symmetric matrices $\subseteq ([\mathbb{R}]_n)^+$. Then for any a, b ε Sym(n) form the matrix c which has a and b as blocks on the diagonal, c = a + b. Now c ε Sym(2n) and $tr(c^2) = tr(a^2) + tr(b^2)$. Also for any symmetric matrix d we know that $tr(d^2)$ is the sum of squares of all of its entries, so $tr(d^2) = 0$ if and only if d = 0. Thus since $a^{\cdot 2} = \frac{1}{2}(a^2 + a^2) = a^2$, and also $b^{\cdot 2} = b^2$, we get $a^{\cdot 2} + b^{\cdot 2} = 0$ if and only if $tr(c^2) = 0$, which implies that $tr(a^2)$ and $tr(b^2)$ are both zero. Consequently a = b = 0.

Because all the Jordan algebras we introduce in our study of variance components will be contained in Sym(n), and hence will be formally real, we get semisimplicity for them by means of a result which forms Exercise 5, p. 205, of Jacobson [1968]:

<u>Theorem 8.3.1.</u> Any finite dimensional formally real Jordan algebra over $\mathbb{R}$ is semisimple, and hence has an identity element and is a direct sum of simple ideals.

We need now one more definition before presenting the classification of our simple Jordan algebras, and this is the idea of involution. We say a linear map a $\rightarrow$ a* on an algebra (associative or non-associative) is an <u>involution</u> if:

[1] $(ab)^* = (b^*)(a^*)$

[2] $(a^*)^* = a$,

for all elements a and b in the algebra. Two examples of involution are: first, complex conjugation in $\mathbf{C}$, and second the transpose in $[\mathbf{R}]_n$. Extending this second example we can transfer an algebra involution to one on the matrices over the algebra as follows. Let $\mathscr{A}$ be an algebra (associative or non-associative) with involution $a \to a^*$, and consider the matrix algebra $[\mathscr{A}]_n$. For $b \in [\mathscr{A}]_n$, $b = [b_{ij}]$ we let

$$b \to b^* = [(b_{ij})^*]'.$$

Then as an example of this induced involution, consider conjugation in the quaternions over a field $\mathscr{F}$. Here $q \in \mathscr{Q}(\mathscr{F})$, $q = a + bi + cj + dk$, implies that $q^* = a - bi - cj - dk$, and one checks that this induces an involution $q \to q^*$ in the matrices over $\mathscr{Q}(\mathscr{F})$, $= [\mathscr{Q}(\mathscr{F})]_n$.

An important requirement we make now of algebras with involution is that all our structures of interest respect the involution: ideals $\mathscr{I}$ are assumed to be such that $\mathscr{I}^* \subseteq \mathscr{I}$, for example. Thus we will call an associative algebra <u>with involution</u> a <u>simple algebra</u> if it has an identity element (so that multiplication is not trivial) and if the only ideals left invariant by the involution (that is, $\mathscr{I}^* \subseteq \mathscr{I}$), are zero and the algebra itself. One can then show that the algebra with involution is simple if and only if it is either simple in the usual sense or is a direct sum of two simple ideals which are exchanged by the involution.

For any matrix algebra $[\mathscr{A}]_n$ with an involution induced from that of $\mathscr{A}$ let's agree to call $a \in [\mathscr{A}]_n$ a <u>symmetric element</u> if $a = a^*$. With this, we can finally state our result on simple Jordan algebras:

Theorem 8.3.2. Every special formally real simple Jordan algebra over $\mathbf{R}$ is isomorphic to one of the following:

[1] $\mathbf{R}$, the real numbers,

[2] $\mathbf{C}$, the complex numbers,

[3] the Jordan algebra of a positive definite symmetric bilinear form f on a vector space $\mathcal{B}$, so $\mathcal{J} = \mathbf{R} \oplus \mathcal{B}$ (see Chapter 7 for this construction),

[4] Sym(n) = the real symmetric matrices, for $n \geq 3$,

[5] the space of all complex matrices $[\mathbf{C}]_n$, for $n \geq 3$, that are symmetric under the involution induced by complex conjugation,

[6] the space of elements of $[\mathcal{Q}(\mathbf{R})]_n$, for $n \geq 3$, that are symmetric under the involution induced by conjugation in $\mathcal{Q}(\mathbf{R})$, the real quaternions.

We will refer to this result as the Second Structure Theorem for Jordan algebras. It is the version of the Theorem of Jordan, von Neumann and Wigner that applies to special Jordan algebras (see Jacobson [1968, p. 205]). It can also be derived by examining the list of all finite dimensional real (central) simple associative algebras with involution (as given for example in Jacobson [1968, Exercises 2 - 7, pp. 211-212]), and then removing all those for which their Jordan spaces of symmetric elements are not formally real. Our result is evidently new, and the fact that all six cases of the theorem are actually realized for specific Jordan algebras is due to Jacobson (personal communication).

We observe that "$n \geq 3$" is required above for otherwise overlap in the cases would occur. Thus in the real 2×2 symmetric matrices consider the space $\mathcal{B}$ spanned by the elements $a_1 = e_{11} - e_{22}$, and $a_2 = e_{12} + e_{21}$. Then we find

$$a_2 \cdot a_1 = 0 \quad \text{and} \quad (a_1)^{\cdot 2} = (a_2)^{\cdot 2} = e_{11} + e_{22} = 1.$$

Next, for $x \in \mathcal{B}$, $x = \alpha a_1 + \beta a_2$ we find $x^{\cdot 2} = f(x)1$ where $f(x) = \alpha^2 + \beta^2 \in \mathbf{R}$, and it follows that $f(x)$ is a real quadratic form on $\mathcal{B}$, inducing a real symmetric bilinear form $g(x, y)$ on $\mathcal{B}$ by means of the standard polarization identity:

$$g(x, y) = \tfrac{1}{4}f(x + y) - \tfrac{1}{4}f(x - y).$$

(See Hoffman and Kunze [1971, p. 368]). Hence $\mathcal{J} = \mathbf{R} \oplus \mathcal{B}$ is a simple Jordan algebra isomorphic to Sym(2).

8.4 Simple Associative Algebras.

We conclude this chapter with a presentation of the structure of any finite dimensional simple associative algebra over $\mathbf{R}$. The result is considerably easier to state than its opposite number for simple Jordan algebras, and is due to Wedderburn. For complete details and the proof see Curtis and Reiner [1962 or 1981]. We will refer to the result as the Second Structure Theorem for Associative Algebras.

Theorem 8.4.1. Let $\mathcal{A}$ be a real finite dimensional simple associative algebra. Then $\mathcal{A}$ is isomorphic to $[\mathcal{B}]_n$ where $\mathcal{B} = \mathbf{R}$, $\mathbf{C}$, or $\mathcal{Q}(\mathbf{R})$, the real quaternions.

Note that the complete collection of simple Jordan algebras above is slightly richer than the parallel collection of simple associative algebras, in that it includes the Jordan algebra of a p.d. symmetric bilinear form.

Chapter Nine: The Algebraic Structure of Variance Components

9.1 Introduction.

We now begin the return to our statistical origins, by applying the algebraic techniques of the last two chapters to our variance component problem. Thus after showing that the space of optimal kernels is, in the kurtosis zero case, a semisimple Jordan algebra, we can invoke the First and Second Structure Theorems of Chapter 8, and this in turn will have purely statistical consequences as shown in Chapter 10. We also briefly examine when a member of the space of optimal estimates is an invariant UMVUE.

In addition, we examine other useful algebraic objects generated by our original mixed model, in particular the Jordan and associative algebras generated by matrices $M_i = MZ_iZ_i'M$, as well as looking at the connection between the two.

Further, by borrowing wholesale from an area of study in the design of experiments we are able to present a collection of important mixed model examples for which the algebraic structure can be fully calculated. The collection is that of the partially balanced incomplete block designs, and so includes the randomized block and the balanced incomplete block models. Thus we are able to fully generalize an example of Seely [1971], who looked at when every estimable function of the components has an optimal unbiased estimate in the case of a balanced incomplete block design.

9.2 The Structure of the Space of Optimal Kernels.

Recall our statistical setting: we have given the k real symmetric matrices $M_i = MZ_iZ_i'M$, corresponding to the k variance components σ_i^2, with $M_k = M = I_n - \mathcal{P}_X$. The linear span of the M_i we have written as $Sp(\sigma^2)$, the space of all j-closed elements of $Sp(\sigma^2)$ is $\mathcal{B}$, and the space of all elements of $Sp(\sigma^2)$ which are both j-closed and s-closed is denoted by $\mathcal{B}^+$, calling this last the space of optimal kernels. For kurtosis $\gamma = 0$ we have seen that $\mathcal{B} = \mathcal{B}^+$, so that all the results presented for $\mathcal{B}$ will carry over to $\mathcal{B}^+$ in this case, which includes the case of normally distributed data.

We begin by proving:

Theorem 9.2.1. $\mathcal{B}$ is a semisimple formally real Jordan algebra.

Proof. We first show that $\mathcal{B}$ is a Jordan algebra. That it is formally real is then immediate, since we have seen that Jordan algebras contained in Sym(n), the space of real symmetric matrices, are necessarily formally real. Semisimplicity then follows from Theorem 8.3.1, but we provide here an alternative proof.

Hence for g, h ε $\mathcal{B}$ we need to prove that gh + hg ε $\mathcal{B}$ as well, so that since $\mathcal{B}$ is a real linear space it follows that $g.h = \frac{1}{2}(gh + hg)$ ε $\mathcal{B}$. To do this begin by assuming that g and h are j-closed, so aga and aha ε $\mathcal{B}$ for all a ε $Sp(\sigma^2)$. In particular we have

$$(c + d)h(c + d) \;\varepsilon\; Sp(\sigma^2)$$

for all c, d ε $Sp(\sigma^2)$, and this occurs if and only if

$$chc + dhc + chd + dhd \;\varepsilon\; Sp(\sigma^2),$$

if and only if

[1] $$dhc + chd \;\varepsilon\; Sp(\sigma^2).$$

Next, as M acts as the identity on $Sp(\sigma^2)$ we know that aM = Ma = a for all a ε $Sp(\sigma^2)$, so ag + ga = agM + Mga ε $Sp(\sigma^2)$. Hence

$$(ag + ga)h(a) + (a)h(ag + ga) \;\varepsilon\; Sp(\sigma^2)$$

where we use $c = ag + ga \in Sp(\sigma^2)$, and $d = a \in Sp(\sigma^2)$, in [1] above. This last occurs if and only if

$$agha + gaha + ahag + agha \in Sp(\sigma^2),$$

which can be written as

$$[(a(gh)a + a(hg)a] + [(g(aha) + (aha)g] \in Sp(\sigma^2),$$

and since $aha \in Sp(\sigma^2)$ the second term in brackets is $\in Sp(\sigma^2)$, so that the first term must also be $\in Sp(\sigma^2)$. Thus $\mathcal{B}$ is closed under the product g.h.

To verify semisimplicity we can argue as follows. Let $\mathcal{R}$ be the Jordan radical of $\mathcal{B}$, so that $\mathcal{R}$ is the unique maximal nilpotent ideal of $\mathcal{B}$, and suppose that $\mathcal{R} \neq 0$ (see Section 7.6). We know that it is then also nil, so that every element of $\mathcal{R}$ is nilpotent. Let $a \in \mathcal{R}$, $a \neq 0$, and suppose p is the smallest positive integer such that $a^{\cdot p} = a^p \neq 0$, while $a^{\cdot 2p} = a^{2p} = 0$. Then

$$0 = tr(a^{2p}) = tr[(a^p)(a^p)],$$

while $a^p \in Sym(n)$, so that necessarily $a^p = 0$, a contradiction, hence only $\mathcal{R} = 0$ is possible. This completes the proof of the theorem. ∎

We observe that the algebraic properties of $\mathcal{B}$ may in fact be carried over to the quadratic estimates themselves by means of the mapping ϕ: $a \rightarrow y'ay$ for $a \in \mathcal{B}$. Thus if $\phi(a) = \phi(b)$ we get $y'(a - b)y = 0$ (a.e.), so in particular

$$E[y'(a - b)y] = \Sigma (\sigma_i^2)tr[(a - b)M_i] = 0,$$

for all σ^2. Hence $tr[(a - b)M_i] = 0$ and $(a - b) \in Sp(\sigma^2)^{\perp}$. But a and $b \in \mathcal{B} \subseteq Sp(\sigma^2)$, so only $a = b$ is possible. Mimicking the Jordan product of $\mathcal{B}$ in the space of of quadratic estimates by letting

$$(y'ay) . (y'by) = y'(a . b)y = \tfrac{1}{2}y'(ab + ba)y$$

we get that ϕ is a Jordan algebra isomorphism. Thus

Theorem 9.2.2. Using the product above, the space of all estimates with j-closed kernels is a formally real semisimple Jordan algebra over $\mathbb{R}$.

Corollary 9.2.3. With $\gamma = 0$, and using the product above, the space of all optimal estimates is a formally real semisimple Jordan algebra.

Other properties of the space of estimates can also be derived from the fact that $\mathscr{B}^+ \subseteq \mathscr{B}$, and that $\mathscr{B}$ is Jordan.

Thus consider the spectral decomposition of $a \in \mathrm{Sym}(n)$:

$$a = \Sigma \lambda_i P_i, \quad 1 \le i \le t,$$

where the λ_i are the eigenvalues of a, $\lambda_i \in \mathbb{R}$ all i, and the P_i are orthogonal idempotents, $P_iP_j = P_jP_i = 0$ for $i \neq j$, $(P_i)^2 = P_i$. Grouping together the positive and negative eigenvalues we write

$$a_+ = \Sigma \lambda_i P_i, \quad \text{for i such that } \lambda_i \ge 0,$$

$$a_- = -(\Sigma \lambda_i P_i), \quad \text{for i such that } \lambda_i < 0,$$

so that

$$a = a_+ - a_-, \quad \text{and} \quad (a_+, a_-) = \mathrm{tr}(a_+a_-) = 0.$$

We call a_+ the positive part of a, a_- the negative part of a, and now show that a_+ and a_- can be expressed as real polynomials in the matrix a. To do this consider the first t positive powers of a:

$$a^j = (\Sigma \lambda_i P_i)^j = \Sigma (\lambda_i)^j P_i, \quad 1 \le j \le t,$$

where we have used the fact that the P_i's are orthogonal idempotents. Using Cramer's Rule (see Hoffman and Kunze [1971, p. 161]) we can solve for the P_i's, getting each as a real polynomial function of a. Thus in any real vector space that contains a and all its powers, we will also find the P_i. In particular, this is true for any associative or Jordan algebra containing the matrix a.

A similar fact holds for the Moore-Penrose g-inverse of a: it is contained in any associative or Jordan algebra which contains a (see Rao and Mitra [1971, p. 61], where the Lagrange-Sylvester interpolation formula for a^+ is given). Also, if a has an inverse we can see by factoring its minimum polynomial that the inverse is expressible as a real polynomial in a.

Putting the above together we can now state:

Theorem 9.2.4. If y'hy is an optimal unbiased estimate, $h \in \mathscr{B}^+$, then the following have their kernels in $\mathscr{B}$: $y'h_+y$, $y'h_-y$, $y'h^+y$, $y'(h^n)y$ for all integers $n > 0$, and $y'(h^{-1})y$, if h^{-1} exists. If $\gamma = 0$ these are all optimal unbiased estimates.

Finally we show a connection between $\mathscr{B}$ and UMVUE's, starting with some further facts about $\mathscr{B}$.

Lemma 9.2.5. Let $\mathscr{G} = \{ bsb \mid b \in \mathscr{B},\ s \in Sp(\sigma^2) \}$. Then $\mathscr{G} \subseteq \mathscr{B}$.

Proof. We want to show that $b_0 \in \mathscr{G}$, $b_0 = bsb$, is such that $a(b_0)a \in Sp(\sigma^2)$ for all $a \in Sp(\sigma^2)$. To this end let

$$e = sba + abs, \text{ for a given } s,\ a \in Sp(\sigma^2).$$

Then since $b \in \mathscr{B}$, it follows that $e \in Sp(\sigma^2)$. Hence also

$$eba + abe \quad \in \quad Sp(\sigma^2),$$

while

$$eba + abe = sbaba + 2absba + ababs.$$

Since

$$s(b)aba + aba(b)s \;\;\varepsilon\;\; Sp(\sigma^2),$$

as $aba \;\varepsilon\; Sp(\sigma^2)$, we see that $absba = a(b_0)a \;\varepsilon\; Sp(\sigma^2)$, as required. ∎

<u>Lemma 9.2.6.</u> Fix $b \;\varepsilon\; \mathscr{B}$. Let $\mathscr{G}_b = \{ bsb \mid s \;\varepsilon\; Sp(\sigma^2) \}$. Then $\mathscr{G}$ and $\mathscr{G}_b$ are Jordan, and $\cup\, \mathscr{G}_b = \mathscr{G}$.

<u>Proof.</u> We want to show

$$bsb(btb)bsb \;\;\varepsilon\;\; \mathscr{G}_b, \text{ for any } s, t \;\varepsilon\; Sp(\sigma^2),$$

as the proof that $\mathscr{G}$ is Jordan is similar, and $\cup\, \mathscr{G}_b = \mathscr{G}$ is a consequence of the definitions. From Lemma 9.2.5 we know $btb \;\varepsilon\; \mathscr{G} \subseteq \mathscr{B} \subseteq Sp(\sigma^2)$, so that

$$b(btb)b \;\;\varepsilon\;\; \mathscr{G} \;\subseteq\; \mathscr{B}$$

after applying the Lemma again. But then

$$s_0 = s[b(btb)b]s \;\;\varepsilon\;\; Sp(\sigma^2),$$

by the definition of $\mathscr{B}$. Thus finally

$$bsb(btb)bsb = b[s[b(btb)b]s]b = b(s_0)b \;\;\varepsilon\;\; \mathscr{G}_b,$$

as required. ∎

<u>Lemma 9.2.7.</u> b and $b^+ \;\varepsilon\; \mathscr{G}_b$, for b^+ the Moore-Penrose inverse of b.

<u>Proof.</u> That $a^+ \;\varepsilon\; \mathscr{J}$ for $a \;\varepsilon\; \mathscr{J}$, $\mathscr{J}$ any Jordan algebra, follows from the discussion leading to Theorem 9.2.4. Hence knowing that $\mathscr{G}_b$ is Jordan, we need only verify that $b \;\varepsilon\; \mathscr{G}_b$. But we know $b^+ \;\varepsilon\; \mathscr{B} \subseteq Sp(\sigma^2)$ for any $b \;\varepsilon\; \mathscr{B}$, so

$$b = b(b^{+})b \ \varepsilon \ \mathcal{G}_b,$$

by the definition of $\mathcal{G}_b$, and we are finished. ∎

To proceed further we need some additional notation. Thus, recalling from Section 5.6, we have $M = KK'$, $K'K = I_t$, $t = r(M)$, so that we can define:

$$Sp_* = \{ \Sigma t_i K' V_i K \mid \text{all real } t_i \}$$

$$Sp_0 = \{ \Sigma t_i V_i \mid \text{all real } t_i \}$$

$$\mathcal{B}_* = \{ b \ \varepsilon \ Sp_* \mid s_* b_* s_* \ \varepsilon \ Sp_*, \text{ for all } s_* \ \varepsilon \ Sp_* \}.$$

Let's observe first that $s \ \varepsilon \ Sp(\sigma^2)$ implies

$$s = Ms_0M = KK's_0KK' = Ks_*K',$$

for $s_0 \ \varepsilon \ Sp_0$, $s_* = K's_0K \ \varepsilon \ Sp_*$. Also note that $b = Kb_*K' \ \varepsilon \ \mathcal{B}$, for $b_* \ \varepsilon \ Sp_*$, if and only if $b_* \ \varepsilon \ \mathcal{B}_*$, and $b_* = K'b_0K \ \varepsilon \ Sp_*$, $b_0 \ \varepsilon \ Sp_0$, is invertible if and only if b_0 is so. Now if the invariant UMVUE $g(y)$ is quadratic, $g(y) = y'by$ (see Section 1.6), then necessarily $b \ \varepsilon \ \mathcal{B}$, and as a partial converse to this we have:

<u>Theorem 9.2.8.</u> Suppose y is normally distributed. In the notation above, if some $b \ \varepsilon \ \mathcal{B}$ is such that b_0 is invertible, then every estimable funtion of the components has an invariant UMVUE.

<u>Proof.</u> If b_0 is invertible then so is $b_* \ \varepsilon \ \mathcal{B}_*$, but then $(b_*).(b_*^{-1}) = I \ \varepsilon \ \mathcal{B}_*$. From this it follows that Sp_* is Jordan and $\mathcal{B}_* = Sp_*$, $\mathcal{B} = Sp(\sigma^2)$. We can apply Theorem 5.6.2 directly to finish the proof. ∎

<u>Corollary 9.2.9.</u> If y is normally distributed with mean zero, and if $\mathcal{B}$ contains a non-singular element, then every estimable function has an I-UMVUE.

9.3 The Two Algebras generated by $Sp(\sigma^2)$.

We next study the Jordan and the associative algebras generated by $Sp(\sigma^2)$, starting with the Jordan case.

Thus consider the sequence of subspaces of Sym(n):

$$\mathscr{L}_0 = Sp(\sigma^2) = \{ \Sigma\, t_i M_i \mid \text{all } i;\ \text{real } t_i \}$$

$$\mathscr{L}_1 = \{ \Sigma\, t_{ij} M_i M_j M_i \mid \text{all } i, j;\ \text{real } t_{ij} \}$$

$$\mathscr{L}_2 = \{ \Sigma\, t_{ij} b_i b_j b_i \mid \text{all } i, j;\ \text{all } b_i, b_j \in \mathscr{L}_1;\ \text{real } t_{ij} \}$$

$$\mathscr{L}_3 = \{ \Sigma\, t_{ij} b_i b_j b_i \mid \text{all } i, j;\ \text{all } b_i, b_j \in \mathscr{L}_2;\ \text{real } t_{ij} \}$$

. . .

By the finite dimensionality of Sym(n) (see also Chapter 7, Section 7.6) the chain

$$Sp(\sigma^2) = \mathscr{L}_0 \subseteq \mathscr{L}_1 \subseteq \mathscr{L}_2 \subseteq \mathscr{L}_3 \subseteq \ldots$$

must become stationary at some integer p, say, with $\mathscr{L}_{p-1} \subseteq \mathscr{L}_p = \mathscr{L}_{p+1} = \mathscr{L}_{p+2} \ldots$. We can now prove

Theorem 9.3.1. $\mathscr{L}_p$ as above is a formally real semisimple Jordan algebra, and is the smallest Jordan algebra in Sym(n) which contains $Sp(\sigma^2)$.

Proof. We need only show that $\mathscr{L}_p$ is Jordan, for the other properties will then follow from the fact that any Jordan algebra $\subseteq$ Sym(n) is necessarily formally real, as we have seen, and so is also semisimple (see Section 8.3), while any Jordan algebra containing $Sp(\sigma^2)$ must contain all the products from which the $\mathscr{L}_i$ are formed.

Thus consider any product aba with a and b $\in \mathscr{L}_p$. Then it is true by construction that aba $\in \mathscr{L}_{p+1}$, but $\mathscr{L}_{p+1} = \mathscr{L}_p$, so right away we are done. ∎

Motivated by this result we make the following

<u>Definition.</u> Denote by $\mathcal{Q}$ the smallest Jordan algebra in Sym(n) which contains $Sp(\sigma^2)$; alternatively, call $\mathcal{Q}$ the <u>Jordan algebra generated by $Sp(\sigma^2)$</u> in Sym(n).

We hence have the chain of subspaces:

$$\mathcal{B}^+ \subseteq \mathcal{B} \subseteq Sp(\sigma^2) \subseteq \mathcal{Q} \subseteq Sym(n).$$

Let's look next at the associative algebra generated by $Sp(\sigma^2)$ in $[\mathbb{R}]_n$. We can see immediately that it is the space of all finite sums of the form

$$\alpha(r_i)^{b(i)}(r_j)^{b(j)} \; . \; . \; . \; (r_t)^{b(t)}$$

where $\alpha \in \mathbb{R}$, b(i) etc. are positive integers not all zero, and the r_i, r_j, etc. are members of the set of all the M_i, $1 \le i \le k$. Note that M is the identity element of this algebra, which we refer to via the

<u>Definition.</u> Denote by $\mathcal{A}$ the smallest associative algebra in $[\mathbb{R}]_n$ containing $Sp(\sigma^2)$; alternatively we say $\mathcal{A}$ is the <u>associative algebra generated by $Sp(\sigma^2)$</u> in $[\mathbb{R}]_n$.

We can therefore extend our subspace chain to:

$$\mathcal{B}^+ \subseteq \mathcal{B} \subseteq Sp(\sigma^2) \subseteq \mathcal{Q} \subseteq \mathcal{A} \cap Sym(n) \subseteq \mathcal{A}$$

where we have observed that $\mathcal{Q}$ must reside in the space of symmetric elements of $\mathcal{A}$.

To study the structural nature of $\mathcal{A}$ and its precise relation to $\mathcal{Q}$ we start with

Lemma 9.3.2. If $a \in \text{rad}\,\mathscr{A}$ and $a' = \beta a$ for $\beta \in \mathbb{R}$, then $a = 0$.

Proof. Since we know $\text{rad}\,\mathscr{A}$ is nilpotent, it is also nil, and for the element $a \neq 0$, $a \in \text{rad}\,\mathscr{A}$, there must exist an integer p such that $a^p \neq 0$, while $a^{2p} = 0$. But then

$$0 = \text{tr}(a^{2p}) = \text{tr}(a^p a^p) = \text{tr}[a^p(1/\beta)^p(a')^p] = (1/\beta)^p \text{tr}[a^p(a')^p],$$

for $\beta \neq 0$. Checking the last term of this equation we see that only $a^p = 0$ is possible, which is a contradiction unless $a = 0$. ∎

With this we are in a position to prove:

Theorem 9.3.3. [1] $\mathscr{A}$ is a semisimple associative algebra;

[2] $\mathscr{A} \cap \text{Sym}(n)$ is a formally real semisimple Jordan algebra, and it is the Jordan algebra generated by the M_i and all terms of the form

$$M_i M_j M_k M_\ell + M_\ell M_k M_j M_i, \quad \text{for } i < j < k < \ell.$$

Proof. To verify [1], begin by assuming $b \in \text{rad}\,\mathscr{A}$. Since M is the two-sided identity of $\mathscr{A}$ (check) it follows that the left ideal in $\mathscr{A}$ generated by b', $= \mathscr{A}[b']$ say, is not zero if $b \neq 0$. Since $\mathscr{A}$ is finite dimensional we can use the ring theory fact that if $\mathscr{A}[b']$ is not nilpotent then it must contain a non-zero idempotent element, $e \in \mathscr{A}[b']$, $e^2 = e$ (see Curtis and Reiner [1962, p. 164]). Now $e = ab'$ for some $a \in \mathscr{A}$, and since $\text{rad}\,\mathscr{A}$ is a two-sided ideal we also have e' an idempotent with $e' = ba' \in \text{rad}\,\mathscr{A}$. But every element of $\text{rad}\,\mathscr{A}$ is nilpotent so $e' = 0$, and also $e = 0$. Thus $\mathscr{A}[b']$ must be nilpotent, hence contained in $\text{rad}\,\mathscr{A}$, since $\text{rad}\,\mathscr{A}$ is not only the unique maximal nilpotent ideal of $\mathscr{A}$, but also contains every one-sided nilpotent ideal (see Curtis and Reiner [1962, p. 161]).

Now let $d = b' - b$, so $d \in \text{rad}\,\mathscr{A}$ and $d' = -d \in \text{rad}\,\mathscr{A}$. By the lemma above d must be zero, so $b' = b$. Here one more application of the lemma gives b = 0, and rad $\mathcal{A} = 0$ as required.

To prove [2] we simply note that this is a straightforward re-working of Cohn's Theorem, applied to the M_i, as in Jacobson [1968, p. 8]. ∎

Corollary 9.3.4. If $Sp(\sigma^2) = \{ \Sigma t_i M_i \mid k \le 3 \}$, that is, if $Sp(\sigma^2)$ has four or fewer spanning elements M_i, that is M_1, M_2, M_3, and $M_4 = M$, then $\mathcal{Q} = \mathcal{A} \cap Sym(n)$.

The point of the corollary is that we always have $\mathcal{Q}$ = the Jordan algebra generated by all the M_i and all products of the form $2\{M_i M_j M_k\} = M_i M_j M_k + M_k M_j M_i$, for $i < j < k$, since for a, b, and c $\varepsilon\ \mathcal{Q}$ we know $(a + c)b(a + c)\ \varepsilon\ \mathcal{Q}$, so also abc + cba $\varepsilon\ \mathcal{Q}$. Also, a slightly stronger version of part [2] of the Theorem above, and of Corollary 9.3.4, is possible though less convenient to evaluate for a given $Sp(\sigma^2)$. Thus Cohn's Theorem (Jacobson [1968]) specifically refers to a generating set of elements of $Sp(\sigma^2)$ for its Jordan closure $\mathcal{Q}$, rather than just a spanning set such as the M_i. Hence Corollary 9.3.4 can be strengthened to: if $Sp(\sigma^2)$ conatains a set { a, b, c } of elements which, along with M, generate $\mathcal{Q}$, then $\mathcal{Q} = \mathcal{A} \cap Sym(n)$.

Having looked at the two algebras generated by $Sp(\sigma^2)$, we turn our attention to the inner structure of $Sp(\sigma^2)$ itself.

9.4 Quadratic Ideals in $Sp(\sigma^2)$.

For brevity, and accuracy in the presence of our several algebras, we say that a Jordan subalgebra of $\mathcal{Q}$ is a $\mathcal{Q}$-ideal if it is a Jordan ideal of $\mathcal{Q}$. Also, recall from Chapter 7 that a Jordan subalgebra $\mathcal{C}$ of $\mathcal{Q}$ is a quadratic ideal of $\mathcal{Q}$ if $\mathcal{C}$ is such that cqc $\varepsilon\ \mathcal{C}$ for all q $\varepsilon\ \mathcal{Q}$, c $\varepsilon\ \mathcal{C}$. The first result is an original, simple and useful criterion for quadraticness in $Sp(\sigma^2)$:

Theorem 9.4.1. Let $\mathscr{C}$ be any additive subgroup of $Sp(\sigma^2)$ which is such that $scs \in \mathscr{C}$ for all $c \in \mathscr{C}$, $s \in Sp(\sigma^2)$. Then $\mathscr{C}$ is a quadratic $\mathscr{Q}$-ideal.

Proof. For any $c \in \mathscr{C}$ and any $s, r \in Sp(\sigma^2)$ we know that rcr and scs $\in \mathscr{C}$, by assumption, while

$$(s + r)c(s + r) = scs + scr + rcs + rcr \in \mathscr{C},$$

so that

[1] $\quad scr + rcs \in \mathscr{C}$, for all $c \in \mathscr{C}$ and all $s, r \in Sp(\sigma^2)$.

Using $s = u$, $r = M$ in [1] gives $uc + cu \in \mathscr{C}$, so that

$$r(uc + cu) + (uc + cu)r = ruc + rcu + ucr + cur \in \mathscr{C}.$$

Next let

$$d = s(ruc + cur)s = srucs + scurs$$

so that $d \in \mathscr{C}$. Further, $scu + ucs \in \mathscr{C}$, so consider

$$\begin{aligned} e &= r(scu + ucs)s + s(scu + ucs)r \\ &= rscus + rucs^2 + s^2cur + sucsr, \end{aligned}$$

which using [1] again is seen to be $\in \mathscr{C}$.

Now letting

$$f = r(scu + ucs) + (scu + ucs)r$$

we know $f \in \mathscr{C}$, and so also is $sf + fs$:

$$\begin{aligned} sf + fs &= srscu + scucs + s^2cur + sucsr \\ &\quad + rscus + rucs^2 + scurs + ucsrs \\ &= srscu + ucsrs + d + e \;\in\; \mathscr{C}, \end{aligned}$$

hence

$$g = srscu + ucsrs \in \mathscr{C}, \text{ for all } c \in \mathscr{C}, \text{ and } s, r \in Sp(\sigma^2).$$

Let's further define

$$h = srscx + xcsrs, \quad \text{and} \quad t = csrs + srsc.$$

Putting $u = x$ in g shows that $h \in \mathscr{C}$, while putting $u = M$ in g shows that $t \in \mathscr{C}$. Then we see $xh + hx \in \mathscr{C}$ while

$$xt + tx = xcsrs + xsrsc + csrsx + srscs$$

$$= csrsx + xsrsc + h \quad \in \mathscr{C},$$

so that $csrsx + xsrsc \in \mathscr{C}$, while defining

$$v = u(csrsx + xsrsc)u = ucsrsxu + uxsrscu$$

gives $v \in \mathscr{C}$. Now $g \in \mathscr{C}$ implies $gxu + uxg \in \mathscr{C}$ by [1], and also

$$gxu + uxg = srscuxu + ucsrsxu + uxsrscu + uxucsrs$$

$$= (srs)c(uxu) + (uxu)c(srs) + v \in \mathscr{C},$$

while $v \in \mathscr{C}$, so we finally get

[2] $(srs)c(uxu) + (uxu)c(srs) \in \mathscr{C}$ for all $c \in \mathscr{C}$, and all $s, r, x, u \in Sp(\sigma^2)$.

The point of [2] is that for any $s_1, t_1 \in \mathscr{L}_1$, and any $c \in \mathscr{C}$, we get

$$s_1ct_1 + t_1cs_1 \in \mathscr{C}.$$

Repeating the argument above now, considering this time s, r, x, and u as elements of $\mathscr{L}_1$ will give

$$s_2ct_2 + t_2cs_2 \in \mathscr{C},$$

for any s_2, $t_2 \in \mathcal{L}_2$. Continuing this way will ultimately lead to

$$s_p c t_p + t_p c s_p \in \mathcal{C}, \text{ for any } s_p, t_p \in \mathcal{L}_p = \mathcal{Q},$$

which is exactly the condition that $\mathcal{C}$ be a $\mathcal{Q}$-ideal.

Finally, we show that $\mathcal{C}$ is quadratic. Thus since $M \in \mathcal{Q}$, and since for any $c \in \mathcal{C}$, $q \in Q$ we know that $cqM + Mqc = cq + qc \in \mathcal{C}$, so also

$$c(cq + qc) + (cq + qc)c \in \mathcal{C}$$

which implies

$$(c^2q + qc^2) + 2cqc \in \mathcal{C}.$$

But $c^2 = \frac{1}{2}[c(c)M + M(c)c] \in \mathcal{C}$ so $c^2q + qc^2 \in \mathcal{C}$, and this implies $cqc \in \mathcal{C}$, as required. ∎

Corollary 9.4.2. An additive subgroup $\mathcal{C}$ of $Sp(\sigma^2)$ is a quadratic $\mathcal{Q}$-ideal if and only if $M_i c M_j + M_j c M_i \in \mathcal{C}$, for all i, j, with $1 \le i,j \le k$.

A useful comparison here is with Lemma 6.4.2 where it is stated that $b \in \mathcal{B}$ if and only if $M_i b M_j + M_j b M_i \in Sp(\sigma^2)$ for all i, j, and to see by just how much $\mathcal{B}$ misses being a quadratic ideal we next examine the interior of $\mathcal{B}$.

9.5 Further Properties of the Space of Optimal Kernels.

We consider the interior of $\mathcal{B}$ by first looking at a special subset of it.

Thus let $\mathcal{B} = \mathcal{K}_0$, and inductively define

$$\mathcal{K}_i = \{ b \in \mathcal{K}_{i-1} \mid sbs \in \mathcal{K}_{i-1} \text{ for all } s \in Sp(\sigma^2) \}.$$

Then $\mathcal{B} = \mathcal{K}_0 \supseteq \mathcal{K}_1 \supseteq \mathcal{K}_2 \supseteq \ldots$, so by the usual argument (finite dimensionality) there must exist an integer ℓ such that $\mathcal{H}_{\ell-1} \supseteq \mathcal{H}_\ell = \mathcal{H}_{\ell+1} = \ldots$. Put $\mathcal{B}_* = \mathcal{H}_\ell$. Then we have

Lemma 9.5.1. $\mathcal{B}_*$ is the unique maximal $\mathcal{Q}$-ideal contained in $Sp(\sigma^2)$.

Proof. Using Theorem 9.4.1 it is, by construction, immediate that $\mathcal{B}_*$ is a quadratic $\mathcal{Q}$-ideal: for $b \in \mathcal{B}_*$ we also have $b \in \mathcal{H}_{\ell+1}$, so $sbs \in \mathcal{H}_\ell = \mathcal{B}_*$ for all $s \in Sp(\sigma^2)$.

To prove that $\mathcal{B}_*$ is unique let $\mathcal{C}$ be any $\mathcal{Q}$-ideal contained in $Sp(\sigma^2)$, so $\mathcal{C} \subseteq Sp(\sigma^2)$ and $qcq \in \mathcal{C}$ for all $c \in \mathcal{C}$, and all $q \in \mathcal{Q}$. Then also $scs \in \mathcal{C}$ for all $c \in \mathcal{C}$, and all $s \in Sp(\sigma^2)$, and by definition of $\mathcal{B}$ it follows that $c \in \mathcal{B}$. Further, by the definition of $\mathcal{H}_1$, c is seen to be $\in \mathcal{H}_1$. But then $scs \in \mathcal{C} \subseteq \mathcal{H}_1$, which implies that $c \in \mathcal{H}_2$. Continuing this way leads to $\mathcal{C} \subseteq \mathcal{H}_\ell = \mathcal{B}_*$, and we are done. ∎

Next let $\mathcal{T}$ be the trace product complement of $\mathcal{B}_*$ in $\mathcal{B}$:

$$\mathcal{B} = \mathcal{B}_* + \mathcal{T}, \text{ with } tr(tb) = tr(bt) = 0 \text{ for all } b \in \mathcal{B}_*,\ t \in \mathcal{T}.$$

Using this definition we derive the following technical lemmas.

Lemma 9.5.2. $btb = 0$ for all $b \in \mathcal{B}_*$, $t \in \mathcal{T}$.

Proof. $tr[(btb)(btb)] = tr[b^2tb^2t] = tr[(b^2tb^2)t]$, while $b^2tb^2 \in \mathcal{B}_*$ by Theorem 9.5.1, so $tr[(btb)^2] = 0$. Since btb is symmetric we must have $btb = 0$. ∎

Lemma 9.5.3. $bst = 0$ for all $b \in \mathcal{B}_*$, $s \in Sp(\sigma^2)$, and $t \in \mathcal{T}$.

Proof. As $bt + tb \in \mathcal{B}$ we get $tr[t(bt + tb)] = 0$, but $tr(tbt) = tr(t^2b)$, and also $tr[t(bt + tb)] = 2tr(tbt)$, and $tr(tbt) = 0$. Next let $b_* = sb^2s$, so we know that $b_* \in \mathcal{B}_*$, and $tr[(bst)(tsb)] = tr[t(sb^2)t] = tr[tb_*t] = 0$, by Lemma 9.5.3, while $(bst)' = tsb$, so only $bst = 0$ if possible. ∎

Out next result is actually standard for semisimple algebras:

Lemma 9.5.4. Every ideal of $\mathcal{B}$ has an ideal-direct complement.

Proof. We have to show that for every $\mathcal{C}$, an ideal of $\mathcal{B}$, with $\mathcal{C}$ not necessarily a a $\mathcal{Q}$-ideal, that there exists another ideal $\mathcal{C}'$ of $\mathcal{B}$ such that $\mathcal{B} = \mathcal{C} + \mathcal{C}'$ with

$$c \,.\, c' = \tfrac{1}{2}(cc' + c'c) = 0, \quad \text{for all } c \in \mathcal{C}, \text{ and } c' \in \mathcal{C}'.$$

To begin, since $\mathcal{B}$ is known to be semisimple there are $\mathcal{B}_i$ such that $\mathcal{B} = \Sigma\, \mathcal{B}_i$, with this an ideal-direct sum, and each $\mathcal{B}_i$ a simple Jordan ideal of $\mathcal{B}$, for $i \in I$. Then let $\mathcal{C} \subseteq \mathcal{B}$ be an ideal of $\mathcal{B}$, and let J be the maximal subset of the index set I such that

$$\mathcal{C}_J = \mathcal{C} + \Big(\sum_{i \in J} \mathcal{B}_j \Big)$$

is an ideal-direct sum, with $j \in J$.

Consider now any $\mathcal{B}_{i^*}$, with $i^* \in I$, i^* not in J. Since $\mathcal{B}_{i^*}$ is simple, the intersection $\mathcal{C}_J \cap \mathcal{B}_{i^*}$ is either zero or $\mathcal{B}_{i^*}$. If it is zero, then i^* can be added to to J, but J is assumed maximal so only $\mathcal{C}_J \cap \mathcal{B}_{i^*} = \mathcal{B}_{i^*}$ is possible. Thus no $\mathcal{B}_{i^*}$ is omitted in the sum $\mathcal{C}_J$, which means $\mathcal{C}_J = \mathcal{B}$ and $\mathcal{B}$ is thereby decomposed as

$$\mathcal{B} = \mathcal{C} \oplus (\Sigma\, \mathcal{B}_j).$$

Putting $\mathcal{C}' = \oplus\, \mathcal{B}_j$ completes the proof. ∎

We can use the lemma to advantage in describing $\mathcal{I}$ as follows:

Lemma 9.5.5. [1] $\mathcal{I}$ is the ideal-direct complement of $\mathcal{B}_*$ in $\mathcal{B}$, so that

$$\mathcal{I} = \oplus\, \mathcal{B}_j, \quad j \in J, \qquad \mathcal{B}_* = \oplus\, \mathcal{B}_i, \quad i \in I, \ i \text{ not in } J.$$

[2] $\mathcal{I}$ is a $\mathcal{Q}$-ideal if and only if $\mathcal{I} = 0$, $\mathcal{B} = \mathcal{B}_*$ and $\mathcal{B}$ is a $\mathcal{Q}$-ideal.

Proof. [1] follows directly from Lemma 9.5.4. Part [2] is arrived at by first using Zorn's Lemma to embed $\mathcal{I}$ in a maximal $\mathcal{Q}$-ideal of $\mathcal{B}$, which by Lemma 9.5.2 must then be exactly $\mathcal{B}_*$. But if this is so then $\mathcal{I} \subseteq \mathcal{B}_*$, so only $\mathcal{I} = 0$ is possible. ∎

Corollary 9.5.6. If $\mathcal{B}$ is a $\mathcal{Q}$-ideal then $\mathcal{B} = \oplus \mathcal{B}_i$ (ideal-direct), with each $\mathcal{B}_i$ a simple $\mathcal{Q}$-ideal.

Proof. First note that $\mathcal{B}$ is a $\mathcal{Q}$-ideal if and only if $\mathcal{B} = \mathcal{B}_*$. Then consider the product sb_js for a given j, $b_j \in \mathcal{B}_j$, and any $s \in Sp(\sigma^2)$. We know that $sb_js = \Sigma b_i$ for $b_i \in \mathcal{B}_i$, with $b_ib_\ell + b_\ell b_i = 0$ for $i = \ell$. Hence if $b_t = e_t$ = the idempotent unit element of $\mathcal{B}_t$ then we must have

$$e_t(sb_js) + (sb_js)e_t = b_j.$$

Next, using the proof of Lemma 9.5.2 with $\mathcal{B}_*$ replaced by $\mathcal{B}_j$, $\mathcal{I}$ replaced by $\mathcal{B}_i$ we get $b_isb_j = 0$ for all $i \neq j$. Thus

$$e_t(sb_js) + (sb_js)e_t = b_t = 0, \quad \text{for } j \neq t,$$

so only $sb_js \in \mathcal{B}_j$ is possible. Theorem 9.5.3 now allows us to conclude that each $\mathcal{B}_j$ is a quadratic $\mathcal{Q}$-ideal. ∎

9.6 The Case of $Sp(\sigma^2)$ Commutative.

We finish this chapter on structure by first seeing how the results above behave when it is assumed that $Sp(\sigma^2)$ is commutative, where by this we mean:

$$M_iM_j = M_jM_i, \quad \text{for all i and j.}$$

This case includes that of all balanced data, as can be verified from Anderson et al. [1984], while we note that in our Example 6.6, cases [2] and [3] are commutative but unbalanced. We will then finally consider a collection of mixed model examples which are not commutative but are such that the structure of the associative algebra $\mathscr{A}$ generated by $Sp(\sigma^2)$ can be readily calculated using some already established facts from the theory of the design of experiments.

We begin with an additional result on formally real Jordan algebras, which the reader will note is a familiar friend in the context of real symmetric matrices:

Theorem 9.6.1. The minimum polynomial of every element in a finite dimensional formally real Jordan algebra has only real roots.

Proof. Suppose that $\mathscr{J}$ is formally real and $a \in \mathscr{J}$. Since the minimum polynomial of a is real, we know from basic algebra that any complex roots appearing in its factorization must occur in conjugate pairs. Select one such pair, assuming such a pair exists, and suppose it is $\delta = \alpha + \beta i$, with $\delta^* = \alpha - \beta i$. Then the minimum polynomial p(x) will factor as

$$\begin{aligned} p(x) &= [(x - \delta)(x - \delta^*)]g(x), \quad \text{with } g(a) \neq 0, \\ &= [x^2 - 2\alpha + \alpha^2 + \beta^2]g(x) \\ &= [(x - \alpha)^2 + \beta^2]g(x). \end{aligned}$$

If $\beta \neq 0$, then element a also solves

$$[p(x)/\beta^2] = p_0(x) = [(x - \alpha)^2/\beta^2 + 1]g(x),$$

as well as

$$h(x) = p_0(x)g(x) = [f(x)g(x)]^2 + [g(x)]^2,$$

for

$$f(x) = (x - \alpha)/\beta.$$

Hence we have the equation in $\mathscr{J}$:

$$h(a) = [f(a)g(a)]^2 + [g(a)]^2 = 0,$$

but as $\mathcal{J}$ is formally real we must have $f(a)g(a) = g(a) = 0$, which contradicts our assumption about $g(x)$. Thus $\beta = 0$, and all roots of $p(x)$ are real. ∎

Our result main is now:

Theorem 9.6.2. If $Sp(\sigma^2)$ is commutative, with $M_iM_j = M_jM_i$ for all i, j, then $\mathcal{Q} = \mathcal{A}$, and $\mathcal{A}$ is a direct sum of copies of $\mathbf{R}$. In particular it has a basis over $\mathbf{R}$ consisting of orthogonal idempotents.

Proof. With the M_i's commuting it follows that $\mathcal{Q}$ is associative:

$$a.b = \tfrac{1}{2}(ab + ba) = \tfrac{1}{2}(ab + ab) = ab,$$

hence $\mathcal{Q} = \mathcal{A}$.

We use Theorem 8.4.1 and observe that for $n \geq 2$, the matrix algebras of the theorem are all non-commutative, as are the real quaternions. Hence the only simple ideals that our $\mathcal{Q} = \mathcal{A}$ can have are just copies of $\mathbf{R}$ or $\mathbf{C}$. However, by Theorem 9.6.1, no formally real Jordan algebra can have an element with minimum polynomial having non-zero complex roots, and a matrix element of $\mathcal{Q} = \mathcal{A}$ which is isomorphic to $i = \sqrt{-1} \in \mathbf{C}$ certainly does. Hence the only simple ideal components possible for $\mathcal{Q} = \mathcal{A}$ are just copies of $\mathbf{R}$. ∎

This result also has a more familiar statistical form, namely that every finite set of commuting, real symmetric matrices can be simultaneously diagonalized by an orthogonal matrix, with each of the matrices then expressible as a real linear combination of orthogonal idempotents.

9.7 Examples of Mixed Model Structure Calculations: The Partially Balanced Incomplete Block Designs.

In this section we calculate $\mathcal{A}$, $\mathcal{Q}$, $\mathcal{B}$ and $\mathcal{B}^+$ for two versions of the class of partially balanced incomplete block designs with association schemes. Other mixed model versions of these designs are possible and can be readily resolved using the calculations we present. We note that Seely [1971] considered a very special case of these models, namely the balanced incomplete block design.

We begin with the observation that A. T. James [1957] was evidently the first to make statistical use of some of the algebraic structure that we have been discussing, by introducing the idea of a "relationship algebra". This associative algebra is that generated by the basic set of incidence matrices used to describe certain fixed effects models. Shortly afterwards Mann [1960] took over the idea to study the "algebra of a linear hypothesis" in a fixed effects model. Other workers in this are include Bose and Mesner [1959], Ogawa and Ishii [1965] and Robinson [1970, 1971].

In all these works the design of greatest interest is that of a fixed effects model that can be construed as an "association scheme" or "concordant graph" that in turn defines a set of symmetric matrices. The associative algebra generated by these matrices along with another algebra, the "relationship algebra", is then used to calculate all sums of squares in an ANOVA table for these designs. See Bose and Mesner [1959], or Ogawa and Ishii [1965], for a full statement of this problem.

Here we show how to borrow from the algebra and ideal calculations of these authors to fully resolve the optimal unbiased estimation problem for the entire class of designs they explicate, the partially balanced incomplete block designs (PBIBDs).

Before proceeding we pause to alert the reader to certain apparent difficulties in this literature:

[1] In the important Ogawa and Ishii [1965] the subalgebras listed at (3.18), p. 1821, are not "two-sided ideals of $\mathcal{R}$", but rather ideals of $\mathcal{R}/\{G\}$; also at (3.16) the matrices A_i^* have nowhere been defined, though they are presumably equal to

$$A_u/(\Sigma c_{ut} z_{ut}).$$

[2] In James [1957, p. 998], one cannot in general presume that an idempotent in a factor ring $\mathcal{R}/\mathcal{A}$, for $\mathcal{A}$ an ideal of $\mathcal{R}$, has a pre-image in $\mathcal{R}$ which is also idempotent. That is, idempotents don't generally "lift" to the original ring: see Jacobson [1964, p. 53]. One can prove however that in James [1957] they do lift as required.

[3] In Robinson [1970] the idempotents E_1, E_2, E_3, and E_4 (any of the possible E_4's given) do not form the desired decomposition of the identity inasmuch as their sum is $w^{-1}G$, not I_w.

With these specific notes of caution, we commence.

We begin by taking over completely the notation and early discussion of Ogawa and Ishii [1965]. Thus start with a design that is characterized by blocks, treatments, and plots (see John [1971] for this basic material). We suppose there are v treatments $\phi_1, \phi_2, \ldots, \phi_v$, and a relation among them satisfying the following conditions:

(a) any two treatments are either 1st, 2nd, ..., or mth <u>associates</u>,

(b) each treatment has exactly n_i ith associates, where n_i may depend on the treatment, for $1 \le i \le m$,

(c) for each pair of treatments which are ith associates, there are p^i_{jk} treatments ($1 \le i, j, k \le m$) which are jth associates of the one treatment of the pair and at the same time kth associates of the other.

Next, we define a partially balanced incomplete block design, a PBIBD, by specifying that there are b blocks each containing k experimental units (plots) in such a way that:

(1) each block contains k ($\leq$ v) different treatments,

(2) each treatment occurs in r blocks,

(3) any two treatments which are ith associates occur together in exactly λ_i blocks, $1 \leq i \leq m$.

If m = 1 then the design reduces to that of a balanced incomplete block, and if further k = v, then the design is said to be a randomized complete block.

A suite of algebraic relations connecting the design parameters is given in Ogawa and Ishii [1965, p. 1816]. Thus if we agree that any treatment may be regarded as the 0th associate of itself, we can put

$$n_0 = 1, \quad \lambda_0 = r, \quad p^0_{jk} = \delta_{jk} n_j, \quad p^i_{0k} = p^i_{k0} = \delta_{ik},$$

for δ_{jk} the Kronecker delta, and then get

$$\Sigma\, n_i = v, \quad \sum_k p^i_{jk} = n_j, \quad \Sigma\, n_i \lambda_i = rk.$$

For this entire class of designs the basic fixed-effects model is

$$y = \mu 1_n + \Phi\tau + \Psi\beta + e,$$

for μ the mean of y, 1_n an n-vector of 1's, τ the fixed-effects treatment parameter, β the fixed-effects block parameter, and e the random error vector, with e distributed as $N(0, \sigma^2_e I_n)$. We now assume our mixed model has the general form as above but take μ, τ

and β to be all random with related variances σ_1^2, σ_2^2, and σ_3^2, and $\sigma_4^2 = \sigma_e^2$, with e not necessarily normal, in fact with arbitrary kurtosis. While we do not claim this model is itself of great practical utility, it is still a very useful intermediate, since calculations with it demonstrate our algebraic methods, and since it will also lead quickly to results for the more useful model that keeps μ a fixed, non-random effect. Moreover, other mixed models flowing from the PBIB design can then be readily evaluated, for example the one having μ and τ fixed, β and e random.

In the standard analysis of the original fixed-effects form of the model, several auxiliary matrices are introduced and these will be essential in our study of the mixed model forms of the model. Thus let A_0 be a unit v × v matrix, and call it the matrix representing the 0th association. For each i, $1 \le i \le m$, let A_i be a symmetric v × v matrix such that its (j, k)th element is one if ϕ_j and ϕ_k are ith associates and is zero otherwise. We call these A_i, $0 \le i \le m$, the <u>association matrices</u>, and it can be seen that

$$\Sigma A_i \quad = \quad (1_v)(1_v)' \quad = \quad J_v.$$

Also, the A_i a linearly independent and since $p^i_{jk} = p^i_{kj}$ we have

$$A_j A_k \quad = \quad \Sigma\, p^i_{jk} A_i \quad = \quad \Sigma\, p^i_{kj} A_i \quad = \quad A_k A_j.$$

It is now immediate that the associative algebra generated in Sym(v) by the A_i is commutative, and in the study of PBIBD's it is called the <u>association algebra</u> of the model. Next, using the commutativity, we can find a new set of independent generators for the algebra by simultaneously diagonalizing the A_i, to get matrices $A_u^{\#}$ which are orthogonal and idempotent, with

$$A_u \quad = \quad \Sigma\, z_{ui} A_i^{\#}, \text{ for constants } z_{ui},\ 0 \le u, i \le m$$

$$A_0^{\#} \quad = \quad (1/v) J_v, \qquad \Sigma\, A_u^{\#} \quad = \quad I_v.$$

Continuing with our definitions let the incidence matrix of blocks be Ψ, such that Ψ is $w \times b$, for w = the number of plots, and such that the (i,j)th element is one if the ith unit belongs to the jth block, and is zero otherwise. Write

$$B = \Psi\Psi', \; w \times w.$$

and call this the block relation. Further, let the incidence matrix of treatments be Φ, such that Φ is $w \times v$, and has (i,j)th element one if the jth treatment occurs at the ith unit, and is zero otherwise. Write

$$T_0 = \Phi\Phi', \; w \times w, \quad \text{with} \quad \Phi'\Phi = rI_w,$$

and more generally

$$T_u = \Phi A_u \Phi', \quad w \times w, \text{ for } 1 \le u \le m.$$

Some important facts about the matrices B and the T_u then are:

$$\Sigma\, T_u = J_w, = G, \text{ say}, \qquad T_0 = G - \Sigma\, T_u,$$

$$G^2 = wG, \qquad BG = GB = kG, \qquad B^2 = kB$$

and

$$GT_u = T_u G = rn_u G, \quad \text{for } 0 \le u \le m.$$

Next let N be the incidence matrix of the design, so that N is $v \times b$, and has (i,j)th element equal to one if treatment i occurs in block j, and is zero otherwise. Then $N = \Phi'\Psi$, and it can be shown that

$$NN' = \Sigma\, \lambda_u A_u = \Sigma\, \rho_u A_u^{\#},$$

where the ρ_u are real constants with $\rho_0 = \Sigma\, n_i \lambda_i = rk$. If all the ρ_u are positive then the design is said to be <u>regular</u>. Using the definition of N one can also show

$$T_0 B T_0 = \Phi N N' \Phi' = \Sigma \lambda_u T_u,$$

$$T_u B T_s = \sum_t \sum_k \sum_\ell (\lambda_k p^\ell_{uk} p^t_{\ell s}) T_t,$$

and

$$T_u T_s = \sum_t p^t_{us} T_t.$$

In the regular case one calls the real associative algebra spanned by the set of 4m + 3 linearly independent matrices

$$\mathscr{L} = \{I_w,\ G,\ B,\ T_u,\ T_u B,\ BT_u,\ BT_u B,\ \text{for } 1 \le u \le m\}$$

the relationship algebra of the PBIBD. An alternative set of basis elements will be more useful for our purposes, so let

$$T_u^{\#} = \Phi A_u^{\#} \Phi', \quad \text{for } 1 \le u \le m.$$

Then

$$T_u = \Sigma z_{ui} T_i^{\#}, \qquad T_0 = \Sigma T_u^{\#},$$

$$T_u^{\#} T_s^{\#} = r \delta_{us} T_u^{\#},$$

$$T_u^{\#} B T_w^{\#} = \rho_u \delta_{uw} T_u^{\#},$$

and the relationship algebra is also now seen to be that spanned by the set

$$\mathscr{L}^{\#} = \{I_w,\ G,\ B,\ T_u^{\#},\ T_u^{\#} B,\ BT_u^{\#},\ BT_u^{\#} B,\ \text{for } 1 \le u \le m\}$$

Moreover, since $T_0 = \Sigma T_u^{\#}$ and since $(1/r)T_u^{\#}$ is a set of orthogonal idempotents, we can apply a result of Section 9.2 to conclude that any Jordan algebra containing T_0 also contains all the $T_u^{\#}$. Thus we see that the Jordan algebra generated by

$$\mathscr{L}_0 = \{I_w,\ G,\ B = \Psi\Psi',\ T_0 = \Phi\Phi'\}$$

is exactly that spanned by

$$\mathscr{L}^{\#} = \{I_w,\ G,\ B,\ T_u^{\#},\ T_u^{\#}B,\ BT_u^{\#},\ T_u^{\#}BT_u^{\#},\ \text{for } 1 \le u \le m\}$$

The connection of the above observations with the random model having four components is thus that the Jordan algebra $\mathcal{Q}$, and the associative algebra $\mathscr{A}$, generated $Sp(\sigma^2)$ is respectively, the Jordan and associative algebra generated by $\mathscr{L}_0$, where $Sp(\sigma^2) = \mathscr{L}_0$, and

$$Z_1 = 1_w, \quad Z_2 = \Phi, \quad Z_3 = \Psi, \quad Z_4 = I_w, \quad M = I_w.$$

$$M_1 = (1_w)(1_w)' = G, \qquad M_2 = \Phi\Phi' = T_0,$$

$$M_3 = \Psi\Psi' = B, \qquad M_4 = I_w.$$

It is also interesting to note that as $G = \Sigma T_u$, $\mathcal{Q}$ is generated as well by the set

$$\{ I_w, B = \Psi\Psi', T_0 = \Phi\Phi' \} \subset \mathscr{L}_0.$$

The structure of $\mathscr{A}$ has been completely determined by Ogawa and Ishii [1965], and Robinson [1970], and this we now present. The calculation of $\mathcal{Q}$ will be an immediate consequence of that of $\mathscr{A}$.

Hence consider the factor algebra $\mathscr{A}/\{G\}$. We find that the sets of elements

$$\{T_u^{\#},\ T_u^{\#}B,\ BT_u^{\#},\ T_u^{\#}BT_u^{\#}\} \bmod \{G\}, \quad \text{for } 1 \le u \le m,$$

form m simple ideals, $(\mathscr{A}/\{G\})_u$ say, for $1 \le u \le m$, are real and four-dimensional if ρ_u is > 0, and one-dimensional if $\rho_u = 0$. Thus using Theorem 8.4.1 these ideals are either copies of $[\mathbb{R}]_2$ or $\mathbb{R}$.

Next define

$$V_u^{\#} = [k/r(rk - ru)](T_u^{\#} - k^{-1}BT_u^{\#})(T_u^{\#} - k^{-1}T_u^{\#}B),$$

and check that the $V_u^{\#}$, $1 \le u \le m$, are orthogonal idempotents; similarly for

$$(1/k\rho_u)BT_u^{\#}B, \quad 1 \le u \le m.$$

Next letting

$$e_u = V_u^{\#} + (1/k\rho_u)BT_u^{\#}B, \quad 1 \le u \le m,$$

we find that each e_u maps onto the identity element of the ideal $(\mathscr{A}/\{G\})_u$. Hence for $a \in \mathscr{A}$, letting $(a)_G$ denote the ideal generated by the image of a in $\mathscr{A}/\{G\}$, we see that

$$\mathscr{A}/\{G\} = \mathscr{H}_1 \oplus \mathscr{H}_2 \oplus (\oplus \mathscr{H}_u)$$

for the $\mathscr{A}/\{G\}$-ideals

$$\mathscr{H}_1 = (k^{-1}B)_G = (B)_G,$$

$$\mathscr{H}_2 = (I_w - (k^{-1}B) + (\Sigma V_u^{\#}))_G,$$

$$\mathscr{H}_u = (e_u)_G = (\mathscr{A}/\{G\})_u$$

Finally since we know that $\mathscr{A}$ must be semisimple, as it is generated by symmetric matrices, we see that the ideal $\{G\}$ in $\mathscr{A}$ has an ideal-direct complement, isomorphic to $\mathscr{A}/\{G\}$, and we have just calculated this last algebra. Hence in summary we find $\mathscr{A}$ is isomorphic to a sum of three copies of $\mathbb{R}$ plus a sum of m copies of either $\mathbb{R}$ or $[\mathbb{R}]_2$, and these are all the simple Wedderburn-Artin components of $\mathscr{A}$. Also,

$$I_w = w^{-1}G + (k^{-1}B - w^{-1}G) + \Sigma V_u^{\#} + (I_w - k^{-1}B - \Sigma V_u^{\#})$$

is a decomposition of the identity of $\mathscr{A}$ into orthogonal idempotents. Each of these idempotents, and others, corresponds to one of the standard sums of squares in the analysis of variance for any PBIB design with association scheme (see Ogawa and Ishii [1965]).

Given the above one can go on to check that $\mathcal{Q}$ is direct sum of the Jordan algebras

$$\mathcal{Q}_1 = \{G\}$$

$$\mathcal{Q}_2 = \{k^{-1}B - w^{-}G\}$$

$$\mathcal{Q}_3 = \{I_w - (k^{-1}B) - (\Sigma V_u^{\#})\}$$

and the m algebras isomorphic to

$$\mathcal{Q}_u = \{T_u^{\#}, T_u^{\#}B + BT_u^{\#}, BT_u^{\#}B\} \bmod G$$

Calculating $\mathcal{B}$, the space of optimal unbiased estimates in the kurtosis zero case, is a straightforward computation, and given the ideal structure above we can check if $\mathcal{B}$ is a $\mathcal{Q}$-ideal or not. Similarly for $\mathcal{B}^+$.

In summary, one finds that the only element $b \in Sp(\sigma^2)$ such that $aba \in Sp(\sigma^2)$ for all $a \in Sp(\sigma^2)$ is $b = \alpha G$, for any $\alpha \in \mathbb{R}$, with

$$E(y'by) = \alpha w(w\sigma_1^2 + k\sigma_2^2 + r\sigma_3^2 + \sigma_e^2),$$

since $tr(G) = w$. Thus we see $\mathcal{B} = \{G\}$ and that it is also a $\mathcal{Q}$-ideal. Moreover since $\mathcal{B}^+ \subseteq \mathcal{B}$, it follows that $\mathcal{B}^+$ is either $\mathcal{B}$ or 0. In fact, another calculation shows $\mathcal{B}^+ = \mathcal{B}$. For all these computations we have, in the notation of Section 6.4:

$$N[1] = [1, \ldots, w],$$

$$N[2] = [w + 1, \ldots, w + v],$$

$$N[3] = [w + v + 1, \ldots, w + v + b],$$

$$N[4] = [w + v + b + 1, \ldots, 2w + v + b];$$

$$\Psi'G = r(1_v)(1_w)', \quad \Phi'G = k(1_b)(1_w)', \quad G_2 = w(1_w)(1_w)' = wG;$$

and if $e_{(i;j)} = \text{diag } e_i$, for e_i the ith unit element of $\mathbb{R}_j$, then

$$s_i = Ge_{(i;w)}G \quad \text{for } i \in N[1],$$

$$s_i = \Phi e_{(i;v)}\Phi' \quad \text{for } i \in N[2],$$

$$s_1 = \Psi e_{(i;b)}\Psi' \quad \text{for } i \in N[3],$$

$$s_i = I_w e_{(i;w)} I_w \quad \text{for } i \in N[4].$$

Hence the products $\Sigma\ s_i G s_i$ for $i \in N[j]$ are evaluated as:

$$\sum_{i \in N[1]} Ge_{(i)}GGGe_{(i)}G \;=\; w^3G \;\in\; Sp(\sigma^2)$$

$$\sum_{i \in N[2]} \Phi e_{(i)}\Phi' G\Phi e_{(i)}\Phi' \;=\; vr\Phi\Phi' = vrT_0 \;\in\; Sp(\sigma^2)$$

$$\sum_{i \in N[3]} \Psi e_{(i)}\Psi' G\Psi e_{(i)}\Psi' \;=\; bk\Psi\Psi' \;=\; bkB \;\in\; Sp(\sigma^2)$$

$$\sum_{i \in N[4]} I_w e_{(i)} G e_{(i)} I_w \;=\; I_w \;\in\; Sp(\sigma^2).$$

Turning now to the three component model, that is the one keeping μ fixed, with τ, β, and e random, we convert to the <u>basic reduced form of the model</u> (1.6) using

$$M = I_w - (1/w)G,$$

and now

$$Sp(\sigma^2) = \{\ aM_1 + bM_2 + cM_3 \mid \text{all real } a, b, c\ \}$$

and where

$$M_1 = M\Phi\Phi'M = \Phi\Phi' - (r/w)G$$

$$= T_0 - (r/w)G, \quad \text{with} \quad tr(M_1) = w - r,$$

$$M_2 = M\Psi\Psi'M = \Psi\Psi' - (k/w)G$$

$$= B - (k/w)G, \quad \text{with} \quad tr(M_2) = w - k,$$

$$M_3 = I_w - (1/w)G = M, \quad \text{with} \quad tr(M) = w - 1.$$

Using the matrix relations developed above, and the structure of $\mathcal{A}$ and $\mathcal{Q}$, we next find that $\mathcal{B}^+ = \mathcal{B} = \{\alpha M \mid \alpha \in \mathbf{R}\}$, $\mathcal{A}_M$ = the associative algebra generated by $Sp(\sigma^2)$ is isomorphic to $\mathcal{A}/\{G\}$, and $\mathcal{Q}_M$ = Jordan algebra generated by $Sp(\sigma^2)$, is isomorphic to $\mathcal{Q}/\{G\}$. Again $\mathcal{B}^+$ is a $\mathcal{Q}$-ideal, and for $b \in \mathcal{B}^+$ we have

$$E(y'by) = \alpha\{(w - r)\sigma_1^2 + (w - k)\sigma_2^2 + (w - 1)\sigma_e^2\}$$

$$= \alpha w(\sigma_1^2 + \sigma_2^2 + \sigma_e^2) - \alpha(r\sigma_1^2 + k\sigma_2^2 + \sigma_e^2).$$

Chapter Ten: Statistical Consequences of the Algebraic Structure Theory.

10.1 Introduction.

We now present some other interwinings of our algebra and statistics, with these results focused in two areas.

In the first area we find that the application of a well-known result on the independence of quadratic forms for normal data provides, via the structure theory, a unique decomposition for any optimal unbiased estimate into statistically independent components, so that a kind of "analysis of variance" is always possible in the space of estimates itself. Thus whereas we began with a mixed model expressing the variance matrix as a sum of "variance components", we will find here conditions for which the optimal unbiased estimates themselves decompose into sums of parts, with each of these being derived from more primitive objects, namely the simple Jordan ideals of $\mathcal{B}$.

The second area is that of non-negative estimation. Here we are able to fully extend a result of Pukelsheim on the question of when an optimal unbiased estimate is non-negative, and more generally, when an estimable function has any non-negative unbiased estimate. Again this proceeds from an ideal decomposition, this time from that of $\mathcal{Q}$.

10.2 The Jordan Decomposition of an Optimal Unbiased Estimate.

We present the first of the two main consequences of the algebraic efforts of Chapters 8 and 9.

Recall that the space of optimal estimates $y'by$, $b \in \mathcal{B}$, is a formally real semisimple Jordan algebra, so that it has a decomposition into an ideal-direct sum of Jordan ideals, each a simple subalgebra, and this is derived from the decomposition of $\mathscr{B} = \Sigma \mathscr{B}_i$, with $\mathscr{B}_i$ the simple ideal components of $\mathscr{B}$.

Theorem 10.2.1. Assume the data vector y is normally distributed. Then every optimal unbiased estimate of the variance components has a decomposition into statistically independent parts:

$$b \in \mathscr{B} \Rightarrow y'by = \Sigma y'b_i y + y'ty,$$

for some $b_i \in \mathscr{B}_i$, $t \in \mathscr{T} \subseteq \mathscr{B}$, and where $y'b_i y$ and $y'b_j y$ are independent for $i \neq j$, and with $y'b_i y$ and $y'ty$ are independent for all i.

Proof. For normal y we know that the data has kurtosis $\gamma = 0$, so that $\mathscr{B}$ is exactly $\mathscr{B}^+$, the space of optimal unbiased estimates.

Next, we can use the proof of Lemma 9.5.3, after replacing $\mathscr{B}_*$ by $\mathscr{B}_i$, and $\mathscr{T}$ by $\mathscr{B}_j$, to conclude that for $b_i \in \mathscr{B}_i$, $b_j \in \mathscr{B}_j$, we have $b_i(s)b_j = 0$ for all $s \in Sp(\sigma^2)$. Using the lemma once more, this time unchanged, gives $b_i(s)t = 0$ for all i, and $t \in \mathscr{T}$, $s \in Sp(\sigma^2)$.

To obtain independence we can argue as follows. From the above we

$$sb_i sb_j s = 0 \quad \text{and} \quad sb_i sts = 0.$$

and since $b \in MbM$, as M is the identity element of $\mathscr{B}$, we get

$$y'by = (My)'b(My), \quad \text{with} \quad E(My) = 0.$$

These facts are sufficient to apply Theorem 9.4.1(a) of Rao and Mitra [1971], and obtain independence as required. ∎

Corollary 10.2.2. Assume y is normally distributed. If $\mathscr{B}$ is a $\mathscr{Q}$-ideal then every optimal unbiased estimate has a decomposition into statistically independent parts as follows:

$$b \in \mathcal{B} \Rightarrow y'by = \Sigma\, y'b_i y, \text{ for } b_i \in \mathcal{B}_i,$$

with $y'b_i y$ and $y'b_j y$ independent for $i \neq j$.

In light of the above let's agree to the following:

Definition. The decomposition of an optimal unbiased estimate as in Theorem 10.2.1 is called the semisimple Jordan decomposition of the estimate, with $y'b_i y$ the ith simple component of $y'by$.

As we have mentioned before it is known from Anderson et al. [1984] that for balanced data $\mathcal{B} = Sp(\sigma^2) = \mathcal{Q}$, and because of the importance of this special case we state this as:

Theorem 10.2.3. Assume y is normally distributed and that the data is balanced. Then every estimable function of the components has an optimal unbiased estimate which has a semisimple Jordan decomposition.

As another example afforded by the decomposition of $\mathcal{B}$ we consider the conditions under which each of the ith simple component of the estimate $y'by$ has a χ^2 distribution. The basic fact needed here is the result appearing in Rao and Mitra [1971] as Theorem 9.2.1. In our situation it reduces to:

Theorem 10.2.4. Let y be normally distributed with mean zero and covariance $s \in Sp(\sigma^2)$. Then $y'ay$ has a $\chi^2(k)$ distribution, with $k = tr(as)$, if and only if

$$sasas = sas.$$

For us s is unknown, and we would generally be interested in checking scaled quadratic estimates of the form

$$g(y) = y'ay, \quad \text{for} \quad a = b/h(\sigma^2),$$

for all σ^2.

The decompositions $\mathcal{Q} = \oplus\, \mathcal{Q}_i$, $\mathcal{Q}_i$ the simple ideal components of $\mathcal{Q}$, and $\mathcal{B} = \oplus\, \mathcal{B}_i$ for the simple ideals $\mathcal{B}_i$, will reduce the study of the equation above to checking it on just the simple components $\mathcal{Q}_i$, or $\mathcal{B}_i$. Thus <u>if $\mathcal{B}$ is a $\mathcal{Q}$-ideal</u> we have seen for $b = \Sigma\, b_i$, $b_i \in \mathcal{B}_i$, that

$$sb_i sb_j s = 0, \quad \text{and} \quad sb_i s \in \mathcal{B}_i, \text{ for all } i, j, i \neq j.$$

so that we need only check

$$sb_i sb_i s \;=\; sb_i s, \quad \text{for all } i.$$

Similarly for $\mathcal{Q}$ and $\mathcal{Q}_i$, so we may examine estimates of the form y'ay as above.

<u>In general then, the semisimple Jordan decomposition of $\mathcal{B}$ allows us to reduce some study of the optimal unbiased estimates to the study of estimates of the form y'by, where b now is an arbitrary element of a simple formally real Jordan algebra, all of which are enumerated in Theorem 8.3.2.</u>

<u>10.3 Non-negative Unbiased Estimation.</u>

The second area of application of the structure theory is that of non-negative estimation, and we begin with the

<u>Definition.</u> An estimable function $p'\sigma^2$ has a <u>non-negative unbiased estimate (n.n.e)</u> if there is a matrix $b \in \text{Sym}(n)$ with b positive semidefinite and y'by unbiased for $p'\sigma^2$.

As usual we write <u>p.s.d.</u> for $b \in \mathrm{Sym}(n)$ with b positive semidefinite. Then we have the following technical lemma (see Pukelsheim [1976]):

<u>Lemma 10.3.1.</u> If b is p.s.d. then so also is $c = \mathcal{P}_{\mathcal{Q}}(b)$, the trace product projection of b onto $\mathcal{Q}$.

<u>Proof.</u> We utilize the decomposition of c into its positive and negative parts, as in Section 9.2: $c = c_+ - c_-$, with c_+ and c_- both p.s.d. Next using $(a_1, a_2) = \mathrm{tr}(a_1 a_2)$ as usual, one can check that if a_1 and a_2 are both p.s.d. then $(a_1, a_2) \geq 0$ so that

$$(b - c, b - c) \geq 0, \quad (b, c_-) \geq 0,$$

$$(c_-, c_-) \geq 0, \quad \text{and} \quad (c_+, c_-) = 0.$$

Hence

$$\begin{aligned}(b - c, b - c) &= (b - c_+, b - c_+) \\ &\qquad + 2(b - c_+, c_-) + (c_-, c_-) \\ &\geq (b - c_+, b - c_+).\end{aligned}$$

However, from the discussion preceding Theorem 9.2.4 we know that $c_+ \in \mathcal{Q}$, since $c \in \mathcal{Q}$, while $c = \mathcal{P}_{\mathcal{Q}}(b)$ minimizes the length $(b - c, b - c)$ over all $c \in \mathcal{Q}$. Hence only $c_- = 0$ and $c = c_+$ is possible. ∎

An immediate use of the decomposition of $\mathcal{B}$ in the study of n.n.e's is as follows. Because $b_i b_j + b_j b_i = 0$ for $b_i \in \mathcal{B}_i$, $b_j \in B_j$, $i \neq j$, we see that b_i and b_j are trace product orthogonal. Then letting $\mathcal{P}_i$ be the projection of Sym(n) onto $\mathcal{B}_i$ and $\mathcal{P}_{\mathcal{B}}$ the projection onto $\mathcal{B}$, we get for any $a \in \mathrm{Sym}(n)$:

$$\mathcal{P}_{\mathcal{B}}(a) = \Sigma\, a_i, \quad \text{for } a_i = \mathcal{P}_i(a) \in \mathcal{B}_i.$$

and a is p.s.d. if and only if all the a_i are p.s.d. Thus we have

Theorem 10.3.2. If an estimable function $p'\sigma^2$ has an n.n.e. $y'ay$ then every $a_i = \mathcal{P}_i(a)$ is p.s.d. and every $y'(a_i)y$ is an n.n.e. If $b \in \mathrm{Sym}(n)$ has $b_i = \mathcal{P}_i(a)$ all p.s.d. then $a = \mathcal{P}_{\mathcal{B}}(b)$ implies that $p'\sigma^2 = E(y'ay)$ has an n.n.e.

Once again, we find that an important estimation question, that of finding n.n.e's, may be specialized to the simple ideal components of $\mathcal{Q}$ and $\mathcal{B}$.

The result concerning non-negative estimation that we propose to extend is, in our notation, the following:

Theorem 10.3.3. (Pukelsheim [1981, Corollary 3]). If $Sp(\sigma^2)$ is commutative, so that $M_iM_j = M_jM_i$ for all i, j, and is Jordan, then there exists an invertible matrix T such that an estimable function $p'\sigma^2$ has an n.n.e if and only if p is such that $p = T\lambda$ for some vector λ with all its components non-negative.

Somewhat ironically we can use our Jordan structure and the Lemma above to drop the requirement that $Sp(\sigma^2)$ be Jordan. By doing so however, we will not be entitled to conclude, as Pukelsheim does in a related result, that the n.n.e. obtained is also optimal unbiased. With this qualification our extension is:

Theorem 10.3.4. If $Sp(\sigma^2)$ is commutative, with $M_iM_j = M_jM_i$ for all i, j, then there exists a matrix T such that an estimable function $p'\sigma^2$ has an n.n.e. if and only if p is such that $p = T\lambda$ for some vector λ with all its components non-negative.

Proof. From Theorem 9.6.2 we know that $Sp(\sigma^2)$ commutative implies that $\mathcal{Q} = \mathcal{A}$ and that $\mathcal{Q}$ has a basis of orthogonal idempotents, which we will denote by P_i for $1 \le i \le t = \dim \mathcal{Q}$. Each M_i can now be expressed as the sum

$$M_i = \Sigma\, t_{ij}P_j, \quad 1 \le i \le t, \quad 1 \le j \le k.$$

Let's now write

$$T = [t_{ij}],$$

so that T is $t \times k$.

For any $\lambda \in \mathbf{R}_t$ consider $b = \Sigma\, \lambda_j P_j$. Then

$$E[y'by] = \xi'\sigma^2,$$

where

$$\xi_i = \mathrm{tr}(bM_i)$$

$$= \mathrm{tr}(\Sigma\, \lambda_j P_j,\ M_i)$$

$$= \Sigma\, \lambda_j \mathrm{tr}(P_j,\ M_i)$$

$$= \Sigma\, \lambda_j \mathrm{tr}(P_j,\ \Sigma\, t_{ir}P_r)$$

$$= \Sigma\, \lambda_j n_j t_{ij},$$

for

$$n_j = \mathrm{tr}(P_j) = \mathrm{rank}(P_j) > 0,$$

so that ξ_i is the ith component of $T(\lambda_*)$, if $(\lambda_*)_i = \lambda_i n_i$. Hence if $p = T\lambda$ then

$$b_* = \Sigma\, (\lambda_i/n_i)P_i$$

implies $y'b_*y$ is unbiased for $p'\sigma^2$. Further, b_* is p.s.d. if and only if

$$\mathrm{tr}(b_*c) \ge 0, \quad \text{for every } c \in \mathcal{Q} \text{ which is p.s.d.},$$

and this holds if and only if

$$\mathrm{tr}(b_* P_i) \ \geq \ 0, \quad \text{for all } i,$$

since the P_i are themselves p.s.d.. But

$$\mathrm{tr}(b_* P_i) \ = \ \lambda_i, \quad \text{for all } i,$$

so $p = T\lambda$ with $\lambda_i \geq 0$ implies $y'b_*y$ is an n.n.e. for $p'\sigma^2$.

Conversely, if $p'\sigma^2$ has an n.n.e, $= y'by$ say, with b p.s.d. then consider the projection $c = \mathcal{P}_{\mathcal{Q}}(b)$ of b onto $\mathcal{Q}$. Using Lemma 10.3.1 it follows that c is p.s.d., so that from the paragraph we see that $c = \Sigma\,\lambda_j P_j$ for some $\lambda \in \mathbb{R}_t$, where $\lambda_j \geq 0$ for all j. Moreover replacing b by its projection onto $\mathcal{Q}$ does not change the expectation, since

$$E(y'by) = \xi'\sigma^2,$$

and the M_i are in $\mathcal{Q}$, so that

$$\xi_i = \mathrm{tr}(bM_i) = \mathrm{tr}[\mathcal{P}_{\mathcal{Q}}(b)M_i] = \mathrm{tr}(cM_i),$$

and

$$E(y'cy) = \ \Sigma\ \mathrm{tr}(cM_i)\sigma_i^2.$$

Thus with $y'by$ unbiased for $p'\sigma^2$, so also is $y'cy$, and $p = T\lambda$ as required. ∎

Finally, we can still recover some optimality in our discussion of n.n.e.'s by using Lemma 10.3.1 once more in:

<u>Theorem 10.3.5.</u> For data y with kurtosis $\gamma = 0$, an estimable function $p'\sigma^2$ has an n.n.e which is also optimal unbiased if and only if its OLS is both j-closed and p.s.d.

<u>Proof.</u> Start by noting that $\mathcal{B}$ is Jordan so that if matrix b is p.s.d. then its projection $c = \mathcal{P}_{\mathcal{B}}(b)$ is also p.s.d., by Lemma 10.3.1. Next, note that if two matrices a_1, and a_2 are $\varepsilon\ \mathrm{Sp}(\sigma^2)$ and are such

that

$$E(y'a_1y) = E(y'a_2y)$$

then necessarily $a_1 = a_2$, since, as usual, $a_1 - a_2 \in Sp(\sigma^2)$ while equal expectations implies $a_1 - a_2 \in Sp(\sigma^2)^\perp$, Then since

$$E(y'cy) = E(y'by)$$

and $b \in \mathcal{B} \subseteq Sp(\sigma^2)$, it follows that $y'by$ is the OLS of $p'\sigma^2$. Thus b is both p.s.d. and j-closed, as required. ∎

Concluding Remarks

Having reached the end of our travels through the wilderness, statistical, algebraic and otherwise, we briefly summarize what has been accomplished and what journeys lie ahead.

First, we have solved the optimal unbiased estimation problem, as we have con-strued it, in Chapter Six. As a prelude to this solution, we also have provided a fully rigorous extension of the results of Seely and Zyskind on the general Gauss-Markov theorem, an extension which allows for the variance matrix to itself be a function of the mean vector that is to be estimated. This effort has mostly meant keeping careful watch on certain variable sets of measure zero, and on the possible constraints on the mean vector.

The other main result, viewed collectively, is that many of the important variance component questions can be studied locally, that is, within a certain family of special formally real simple Jordan algebras, with the complete family being given by Theorem 8.3.2.

As we noted in the introduction to Chapter Eight, to bring this program of localization to fruition one requires a method for calculating these simple components from the set of generating matrices in the given mixed model, and this is an algebra research problem itself. Hence while we may use this structure for obtaining new facts, such as Theorem 10.3.4 on the Jordan decomposition of any optimal estimate into independent quadratic estimates, and may also use the structure for insight generally, the problem remains of getting the simple Jordan components B_i in each practical problem of interest. We hope however, that our demonstration of the power and simplicity (semisimplicity?) of these algebraic methods will encourage this research, at least for the most frequently used mixed models, since for these the generators M_i are usually few in number.

Finally, three other directions may be followed. The first is that of extending our work, at all points, to the non-zero kurtosis case, that is, of seeing how the reductive structure results interact with the study of estimates that are s-closed in addition to being j-closed. The second direction assumes that the basic model does not take $Z_k = I_n$, and pursues a slightly more general version of variance components analysis. Note however that by Theorems 8.2.1 and 8.3.1, both $\mathcal{A}$ and $\mathcal{Q}$ necessarily have identity elements. Lastly, there is the problem of seeing how any of our results will simplify or complete the study of biased estimation of the variance components, that is, how minimum mean square error can be attained. The Universe containing these particular forest wildernesses beckons us onward.

References

Anderson, R. D. [1978]. Studies on the estimation of variance components. Ph.D. Thesis, Cornell University.

Anderson, R. D., Henderson, H. V., Pukelsheim, F., and Searle, S. R. [1984]. Best estimation of variance components from balanced data, with arbitrary kurtosis. Math. Operationsforsch. u. Statis., ser. statist. [15], 163-176.

Anderson, T. W.[1948]. On the theory of testing serial correlation. Skand. Aktuarietidskr. [31], 88-116.

Bickel, P. J., and Doksum, K. A. [1977]. Mathematical Statistics. Holden-Day, San Francisco.

Bose, R. C., and Mesner, D. M. [1959]. On linear associative algebras corresponding to association schemes of partially balanced designs. Annals of Math. Stat. [30], 21-38.

Curtis, C. W., and Reiner, I. [1966]. Representation Theory of Finite Groups and Associative Algebras, Second edition. Wiley, New York.

Curtis, C. W., and Reiner, I. [1981]. Methods of Representation Theory with Applications to Finite Groups and Orders. Wiley, New York.

Drygas, H. [1977]. Best quadratic unbiased estimation in variance-covariance component models. Math. Operationsforsch. u. Statis., ser. statist. [8], 211-231.

Drygas, H. [1980]. Hsu's theorem in variance component models. In:

Mathematical Statistics, Banach Center Publications, vol. 6, PWN - Polish Scientific Publishers, 95-107.

Drygas, H. [1985]. Estimation without invariance and Hsu's theorem in variance component models. In: Contributions to Econometrics and Statistics Today, edited by H. Schneeweiss and H. Strecker, Springer-Verlag, 56-69.

Eaton, M. L. [1970]. Gauss-Markov estimation for multivariate linear models: a coordinate free approach. Annals of Math. Stat. [41], 528-538.

Eaton, M. L. [1983]. Multivariate Statistics: a Vector Space Approach. Wiley, New York.

Elbassiouni, M. Y. [1983]. On the existence of explicit restricted maximum likelihood estimators in multivariate normal models. Sankhya, Series B, 303-305.

Farrell, R. H. [1976]. Techniques of Multivariate Calculation. Springer-Verlag, New York.

Farrell, R. H. [1985]. Multivariate Calculation: Use of the Continuous Groups. Springer-Verlag.

Goldman, A. J., and Zelen, M. [1964]. Weak generalized inverses and minimum variance linear unbiased estimation. J. Res. Nat. Bur. Stand. [68B], 151-172.

Graham, A. [1981]. Kronecker Products and Matrix Calculus: with Applications. Wiley (Halsted Press), New York.

Harville, D. A. [1981]. Unbiased and minimum-variance unbiased estimation of estimable functions for fixed linear models with artibrary covariance structure. Annals of Stat. [9], 633-637.

Henderson, H. V., and Searle, S. R. [1979]. The vec-permutation matrix, the vec operator and Kronecker products: a review. Paper BU-645-M in the Biometrics Unit, Cornell University.

Herstein, I. N. [1968]. Noncommutative Rings. Carus Mathematical Monograph No. 15, The Mathematical Association of America.

Herstein, I. N. [1969]. Topics in Ring Theory. University of Chicago Press.

Hoffman, K., and Kunze, R. [1971]. Linear Algebra. Prentice-Hall, New Jersey.

Jacobson, N. [1964]. Structure of Rings. American Mathematical Society, Rhode Island.

Jacobson, N. [1968]. Structure and Representations of Jordan Algebras. American Mathematical Society, Rhode Island.

James. A. T. [1957]. The relationship algebra of an experimental design. Annals of Math. Stat. [28], 993-1002.

John, P. W. M. [1971]. Statistical Design and Analysis of Experiments. Macmillan, New York.

Khatri, C. G. [1978]. Minimum variance quadratic unbiased estimate of variance under ANOVA model. Gujarat Statistical Review [V], 33-41.

Kelley, J. L. [1955]. General Topology. D. Van Nostrand, Princeton, New Jersey.

Kendall, M., and Stuart, A. [1977]. The Advanced Theory of Statistics, Volume 1, Fourth edition, Macmillan, New York.

Khuri, A. I., and Sahai, H. [1985]. Variance components analysis: A selective literature survey. Int. Statis. Rev. [53], 279-300.

Kleefe, J. [1977]. Optimal estimation of variance components - a survey. Sankhya, Series B, 211-244.

Kruskal, W. [1968]. When are Gauss-Markov and least squares estimators identical? A coordinate-free approach. Annals of Math. Stat. [39], 70-75.

Lang, S. [1965]. Algebra. Addison-Wesley, Reading, MA.

LaMotte, L. R. [1973a]. On non-negative quadratic unbiased estimation of variance components. J. Amer. Stat. Assoc. [68], 728-730. 728-730.

LaMotte, L. R. [1973b]. Quadratic estimation of variance components. Biometrics [29], 311-330.

LaMotte, L. R. [1976]. Invariant quadratic estimators in the random, one-way ANOVA model. Biometrics [32], 793-804.

Lehmann, E. L. [1959]. Testing Statistical Hypotheses. Wiley, New York.

Lehmann, E. L., and Scheffé, H. [1950]. Completeness, similar regions and unbiased estimation - Part 1. Sankhya [10], 305-340.

Mann, H. B. [1960]. The algebra of a linear hypothesis. Annals of Math. Stat. [31], 1-15.

McCoy, N. H. [1964]. The Theory of Rings. Macmillan, Toronto, Ontario.

Mitra, S. K. [1971]. Another look at Rao's MINQUE of variance components. Bull. Inst. Internat. Statist. [44], 279-283.

Morin-Wahhab, D. [1985]. Moments of a ratio of two quadratic forms. Comm. Statist. - Theor. Meth. 14(2), 499-508.

Ogawa, J., and Ishii, G. [1965]. The relationship algebra and the analysis of variance of a partially balanced incomplete block design. Annals of Math. Stat. [36], 1815-1828.

Olsen, A., Seely, J., and Birkes, D. [1976]. Invariant quadratic unbiased estimation for two variance components. Annals of Stat. [4], 878-890.

Pukelsheim, F. [1976]. Estimating variance components in linear models. J. of Multi. Analysis [6], 626-629.

Pukelsheim, F. [1981]. On the existence of unbiased nonnegative estimates of variance coveriance components. Annals of Stat. [9], 293-299.

Rao, C. R. [1967]. Least squares theory using an estimated dispersion matrix and its application to measurement of signals. Proc. Fifth Berkeley Symposium on Mathematical Statistics and Probability, vol. 1, 355-372.

Rao, C. R. [1971a]. Estimation of variance and covariance components - MINQUE theory. J. of Multi. Analysis [1], 257-275.

Rao, C. R. [1971b]. Minimum variance quadratic unbiased estimation of variance components. J. of Multi. Analysis [1], 445-456.

Rao, C. R. [1973a]. Linear Statistical Inference and its Applications. Second edition, Wiley, New York.

Rao, C. R. [1973b]. Representations of best linear unibased estimators in the Gauss-Markov model with a singular dispersion matrix. J. of Multi. Analysis [3], 276-292.

Rao, C. R., and Mitra, S. K. [1971]. Generalized Inverse of Matrices and its Applications. Wiley, New York.

Robinson, J. [1970]. On relationship algebras of incomplete block designs. Annals of Math. Stat. [41], 648-650.

Robinson, J. [1971]. Correction to "On relationship algebras of incomplete block designs". Annals of Math. Stat. [42], 842.

Rogers, G. S. [1980]. Matrix Derivatives. Dekker, New York.

Sahai, H. [1979]. A bibiliography on variance components. Internat. Statist. Rev. [47], 177-222.

Sahai, H., Khuri, A. I., and Kapadia, C. H. [1985]. A second bibliography on variance components. Commun. Statist. - Theor. Meth. 14(1), 63-115.

Searle, S. R. [1971]. Linear Models. Wiley, New York.

Searle, S. R. [1979]. Notes on variance component estimation: A detailed account of maximum likelihood and kindred methodology. Paper BU-673-M in the Biometrics Unit, Cornell University.

Searle, S. R. [1984]. Cell means formulation of mixed models in the analysis of variance. Paper BU-862-M in the Biometrics Unit, Cornell University.

Seely, J. [1971]. Quadratic subspaces and completeness. Annals of Math. Stat. [42], 710-721.

Seely, J. [1972]. Completeness for a family of multivariate normal distributions. Annals of Math. Stat. [43], 1644-1647.

Seely, J. [1977]. Minimal sufficient statistics and completeness for multivariate normal families. Sankhya, Series A, 170-185.

Seely, J., and El-Bassiouni, Y. [1983]. Applying Wald's variance component test. Annals of Stat. [11], 197-201.

Seely, J., and Zyskind, G. [1971]. Linear spaces and minimum variance unbiased estimation. Annals of Math. Stat. [42], 691-703.

Szatrowski, T. H. [1980]. Necessary and sufficient conditions for explicit solutions in the multivariate normal estimation problem for patterned means and covariances. Annals of Stat. [8], 802-810.

Szatrowski, T. H., and Miller. J. J. [1980]. Explicit maximum likelihood estimates from balanced data in the mixed model of the analysis of variance. Annals of Stat. [8], 811-819.

Taylor, S. J. [1966]. Introduction to Measure and Integration. Cambridge University Press.

van der Waerden, B. L. [1970]. Algebra. Frederick Ungar, New York.

Zyskind, G. [1967]. On canonical forms, non-negative covariance matrices and best and simple least squares linear estimators in linear models. Annals of Math. Stat. [38], 1092-1109.